The Mathematica® Programmer II

The MATHEMATICA® Programmer II

ROMAN E. MAEDER

Department of Computer Science
Swiss Federal Institute of Technology
Zürich, Switzerland

ACADEMIC PRESS

San Diego   Boston   New York   London   Sydney   Tokyo   Toronto

Find Us on the Web! http://www.apnet.com

This book is printed on acid-free paper. ∞

Copyright © 1996 by ACADEMIC PRESS, INC.

All Rights Reserved.
No part of this publication may be reproduced or transmitted in any form or by any means, electronic or mechanical, including photocopy, recording, or any information storage and retrieval system, without permission in writing from the publisher.

Academic Press, Inc.
A Division of Harcourt Brace & Company
525 B Street, Suite 1900, San Diego, California 92101-4495

United Kingdom Edition published by
Academic Press Limited
24-28 Oval Road, London NW1 7DX

Library of Congress Cataloging-in-Publication Data

Maeder, Roman.
    The mathematica programmer II / by Roman E. Maeder.
          p.   cm.
    Includes bibliographical references and indexes.
    ISBN 0-12-464992-0
    1. Mathematica (Computer program language)   I. Title.
    QA76.73.M29M34   1996
    510'.285'53--dc20   96-14916
                                                        CIP

PRINTED IN THE UNITED STATES OF AMERICA
96 97 98 99 00 01 EB 9 8 7 6 5 4 3 2 1

Contents

Preface ix

About This Book
    Overview xi
    About the Programs xii
    Notation and Terminology xv
    The Mathematica Programmer CD-ROM xv
    The Mathematica Programmer WWW Archive xvi
    Colophon xvii

Part 1: Paradigms of Programming

1 Introduction
    1.1 *Mathematica*'s Programming Language 5
    1.2 Pattern Matching and Term Rewriting 8
    1.3 Programming Styles 10
    1.4 Program Organization 18

2 Logic Programming
    2.1 The Ingredients of Logic Programs 23
    2.2 A PROLOG Interpreter for *Mathematica* 30
    2.3 Lists in PROLOG 44
    2.4 Backtracking 47
    2.5 Deduction 51
    2.6 Conclusions 55

3 Higher-Order Functions
    3.1 Introduction 59
    3.2 The Functional Features of *Mathematica* 59
    3.3 Functions as Data 61
    3.4 Fixed Points of Higher-Order Functions 65

4 Combinators
    4.1 Introduction 75
    4.2 Combinatory Algebras 75
    4.3 Combinatory Abstraction 78
    4.4 Converting Functions to Combinators 81
    4.5 Applications 84

5 Turing Machines
    5.1 Introduction 89

5.2 A Turing Machine Simulator 91
5.3 Assembly Programming 95
5.4 Recursive Functions 98
5.5 Optimization 106
5.6 Conclusions 109
5.7 The Complete Code of TuringMacros.m 113

Part 2: Visualization

6 Animated Algorithms
6.1 Three Standard Sorting Algorithms 121
6.2 Sorting in Action 123
6.3 Asymptotic Behavior 130
6.4 Conclusions 132

7 Function Iteration and Chaos
7.1 Function Iteration 137
7.2 Bifurcations 142
7.3 The Final-State Diagram 149
7.4 The Ingredients of Chaos 152
7.5 Super-Attractive Orbits 157
7.6 Conclusions 162

8 Fractional Brownian Motion
8.1 Introduction 165
8.2 Random Additions 167
8.3 Fourier Synthesis 176
8.4 Random Faults 180
8.5 Analysis of fBm Data 184

9 Uniform Polyhedra
9.1 Introduction 187
9.2 Uniform Construction 189
9.3 Data Structures 197
9.4 Rendering 200
9.5 Auxiliary Programs 205

10 The Stellated Icosahedra
10.1 Introduction 211
10.2 Stellations of the Icosahedron 212
10.3 Rendering 219
10.4 Discussion 224
10.5 The Complete Code of Icosahedra.m 227

11 Ray Tracing
    11.1 A Data Type for Surfaces 237
    11.2 Photorealistic Rendering 240
    11.3 Converting *Mathematica* Graphics 241
    11.4 Sample Images 245
    11.5 Stereo Pairs 249
    11.6 The Complete Code of POVray.m 251

12 Single-Image Stereograms
    12.1 Introduction 257
    12.2 The Classic SIRDS in *Mathematica* 257
    12.3 Designing Good Images 262
    12.4 Exact Stereograms 264
    12.5 Interface to External SIS Generators 267
    12.6 The Complete Code of SIS.m 274

Appendices

References 281

Index of Programs 285

Index 287

Preface

Mathematica's programming language is an important addition to a comprehensive symbolic computation system. To the novice, it may seem overwhelming, with over 1000 built-in functions. As we shall show in Chapter 1, it nevertheless has a coherent style and it is easy to learn. The significance of *Mathematica*'s programming language can also be judged from the fact that it is used for teaching programming at a growing number of universities. One of the most rewarding features is that a clarifying picture and other means of explaining how programs work are never far away. One has the whole power of *Mathematica* at one's disposal.

The large number of ways to solve some given problem makes it necessary to point out the easy way of doing things, especially to users who have been familiar with one of the traditional programming languages. The designers' view was expressed in my first book, *Programming in Mathematica*. My goal was to explain the ideas behind the language and to develop useful example programs. This strategy is continued in an ongoing series of articles in *The Mathematica Journal*, entitled *The Mathematica Programmer*. The title of this series is now also the title of this book series. A first installment of these articles was published in the first volume. Now, another set of articles has accumulated and is part of this Volume II. The articles can be divided roughly into two kinds: explanations of fundamental programming paradigms and applications, this time mostly visualizations. This distinction is reflected in the two parts that make up this book. Besides the articles from the journal, and one article from *Mathematica in Education*, I have included new material, such as Chapter 1, which has been updated from a similar introduction in the first volume. The material from the articles has been expanded and all program listings are now included.

The first articles appeared before Version 2.0 of *Mathematica*, and all articles appeared before Version 3. Because there have been major improvements from Version 1 to Version 3, as well as a few incompatibilities, I took the opportunity to update all programs from both volumes to Version 3. All programs are included in the accompanying CD-ROM.

Pictures are always a good means of explaining things. The beauty of well-chosen graphics illustrations is in itself an important aspect of the otherwise rather dry world of computers. The color insert in this book gives me the chance to show the full range of possibilities. In all cases *Mathematica* delivered the raw data, which was then turned into color illustrations using a variety of techniques, which are the topic of the second part of this book.

The CD-ROM contains all the information relating to this book that is best accessed in electronic form, such as the programs, notebooks with examples, color images, and pointers to resources on the Internet. The information on the CD-ROM can be accessed through a hypertext system.

My thanks go foremost to the editors of *The Mathematica Journal* and *Mathematica in Education*, Troels Petersen and Paul Wellin, respectively. They helped me put my writings into a form worthy of publication. I would like to thank Miller-Freeman, Inc., for giving me permission to include the articles from *The Mathematica Journal* in this book, and Springer-Verlag/TELOS for doing likewise with the article that appeared first in *Mathematica in Education* (now called

Mathematica in Education and Research). Chuck Glaser from Academic Press encouraged me to publish this second volume.

A number of people contributed to particular subject matters. Robert Marti helped me develop the logic interpreter, and Georgios Grivas and Stephan Zahner wrote the first version of the unifier. Zvi Har'El made many helpful comments on a draft of Chapter 9. The visualization chapters benefit from the easy availability of good ray-tracing and stereogram programs. My thanks go foremost to the POVRAY team and to Craig E. Kolb for writing the ray tracers, and to Pascal Massimino for his RAYSIS. The programs work very well. Special recognition to Jef Poskanzer for developing the PBM package, which successfully tackles the jungle of graphic file formats. John Bradley's program XV was most useful for looking at the graphics on my workstation screen. More information about all of these programs is included on the *Mathematica Programmer* CD-ROM. Plate 13 was produced by F. Bachmann, ETH Zurich, using my ray-traced images. M. Braschler helped me prepare the material on the CD-ROM.

Herrliberg, Switzerland   R. E. M.

About This Book

This section gives an overview of the contents of the book, describes conventions used, and discusses the electronic resources available to readers of the book.

Overview

Following the tradition established with the first volume [38], *The Mathematica Programmer II* is divided into two parts.

Part 1 discusses different important programming paradigms in greater detail. The introduction in Chapter 1 gives an overview; it includes also those topics that were treated in the first volume of this book.

Chapter 2 is about logic programming. This programming style is not built into *Mathematica*, but we can easily develop an interpreter for it. Logic programs differ conceptually from any other kind of program. They are well suited for a number of tasks including theorem proving, backtracking, and exhaustive search.

In Chapter 3 we take a closer look at functional programming and at higher-order functions. Some of the material is rather advanced, but we give many practical examples that illustrate the usefulness of these ideas. Even a bit more technical is Chapter 4. It treats an interesting approach to defining functions without the use of variables. These ideas are easily realized in *Mathematica*.

The most important theoretical model of a computer is the Turing machine. We use it in Chapter 5 to show how to construct interpreters and to discuss compilation methods for programming languages.

Part 2 of the book is entirely devoted to visualization. We use *Mathematica*'s excellent graphic capabilities to develop images for various applications, from geometry to the study of algorithms. For a number of computer-graphics methods not supported directly, we develop tools that convert *Mathematica* graphics into input files for external renderers. In this way, we use the symbolic and numerical capabilities to generate data for other programs.

Chapter 6 shows how we can visualize the performance of three standard sorting algorithms. A simple modification in an auxiliary function allows us to display the sequence of steps used to sort lists of data.

Chaos and fractals have become a fashionable subject, partly due to the stunning graphics that can be produced from chaotic and fractal systems. We investigate some chaotic phenomena both symbolically and numerically in Chapter 7, and, of course, develop programs for our own chaotic pictures. Chapter 8 treats a method to create fractal lines, surfaces, or clouds: fractional Brownian motion. It is being used to generate realistic landscapes.

Geometry is the topic of the next two chapters. First, in Chapter 9, we discuss uniform polyhedra and develop ways to visualize them. Then, in Chapter 10, we look at stellated icosahedra. Their generation requires symbolic manipulation, and we use numerical methods to create the pictures.

The imaging method that gives the most realistic pictures is ray tracing. It turns out to be relatively easy to convert any three-dimensional *Mathematica* graphic into input for ray-tracing programs. We develop the necessary tools in Chapter 11. An excursion into ray-traced stereo pairs concludes the chapter.

Of course, we *have* to use *Mathematica* to produce stereograms, those pictures that caused so many people to stare cross-eyed at seemingly meaningless color pictures. We explain how these images work and how we can create them directly in *Mathematica* or with external renderers in Chapter 12.

The appendix contains the bibliographic references. It is followed by an index of all programs and the subject index. An insert with color graphics contains high-resolution images related to the examples in the book.

This book is not an introduction to *Mathematica*. Syntactic details of its programming language are not treated here. *Mathematica* comes with a good manual (*The Mathematica Book* [61]) which explains everything we use here. The reference guide and introductory texts are available on-line in the help browser. If you are serious about programming in *Mathematica*, I recommend to you my book on this subject [40].

About the Programs

All examples were tested with a prerelease of Version 3.0 of *Mathematica*. The "live" calculation sequences in this book were computed on a Sun SPARCstation 5 running SunOS 4.1.4. Listing init.m shows the initialization commands used for our calculations. No attempt has been made to ensure that programs work with earlier versions of *Mathematica*. On the contrary, I updated all programs from Volume 1 and Volume 2 to Version 3.0. The programs are enclosed on the accompanying CD-ROM.

If an output is uninteresting, we often terminate a command by a semicolon to suppress the output.

```
In[1]:= p100 = Expand[(x+1)^100];
```

The output of a command that produces a picture is most often uninteresting; therefore, most graphic commands are terminated by a semicolon. The picture is a side effect, not the output!

```
In[2]:= Plot[ Sin[x + Sin[3x]], {x, 0, 2Pi} ];
```

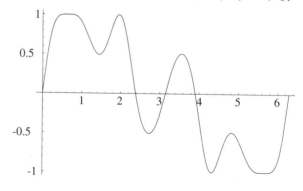

About This Book

```
SetOptions["stdout", PageWidth->60]   (* line width *)
Format[Continuation[_]] := ""         (* no blank lines *)

SeedRandom[10000]                     (* reproducible "random" numbers *)

Off[ General::spell, General::spell1 ]

SetOptions[ ParametricPlot3D, Axes -> None ]

Needs["MathProg`SystemProgramming`"]
SetAllOptions[ ColorOutput -> GrayLevel ]

$DefaultFont = {"Times-Roman", 9.0}

Begin["`Private`"]
Unprotect[Short]
Short[e_] := Short[e, 2]              (* lines are very short *)
Protect[Short]
End[]
Null
```

init.m: *Mathematica* initialization for this book

In the interactive calculations we sometimes use statements of the form (*var = expr*) // Short to perform an assignment and show an outline of the long result it generates. Note the parentheses around the assignment.

$$\text{In[3]:= (p100 = Expand[(x+1)}\wedge\text{100]) // Short}$$
$$\text{Out[3]//Short=}$$
$$1 + 100\,x + 4950\,x^2 + 161700\,x^3 + 3921225\,x^4 + \langle\langle 94\rangle\rangle + 100\,x^{99} + x^{100}$$

The MathProg directory on the CD-ROM contains the programs. You can access the programs directly from the CD-ROM. Simply load the init.m package with a command of the form Get["*full pathname of init.m*"] The pathname is system dependent and may look like this:

Pathname	Operating System
D:\MATHPROG\INIT.M	Windows
Mathematica Programmer:MathProg:init.m	Macintosh
/cdrom/mathprog/MathProg/init.m	Solaris
/mathprog/MathProg/init.m	NeXTStep
/mnt/cdrom/MathProg/init.m	Linux

You can use the frontend's Get File Path menu command to obtain the pathname easily. The effect of reading init.m is to preload all major book packages. The first time you use a function, the corresponding package is read in automatically.

If you use the programs frequently, you should copy the MathProg directory onto your hard disk, into the directory AddOns/Applications (inside your *Mathematica* distribution) or into your home directory on multi-user operating systems. If you copy it there, *Mathematica* will automatically find the packages; no further installation is necessary. To preload the packages, give the command

Get["MathProg`init`"] (or <<MathProg`init`). Please refer to the README file for further information. The packages from Volume 1 can be preloaded by reading MathProg`init1`.

This command preloads all packages assuming the MathProg directory has been properly installed on the hard disk.	`In[4]:= << MathProg`init``

You can see the contexts of all our packages. The packages themselves have not yet been read, however.	`In[5]:= $ContextPath // Short`
	`Out[5]//Short=`
	`{MathProg`Unify`, MathProg`UniformPolyhedra`,`
	` MathProg`TuringRecursive`, <<21>>, Global`, System`}`

When a function is used for the first time, the corresponding package is loaded automatically. (The functions in this example are found in the package IteratedFunctions.m.)	`In[6]:= FindPeriod[LogisticsMap[3.45], MaxIterations->30]` `Out[6]= {0.445968, 0.852428, 0.433992, 0.847468}`

> Files that are not proper packages (for example, DAG.m), as well as packages that redefine system functions (for example, SurfaceGraphics3D.m), cannot be preloaded in this way. Such files must be read explicitly before their functions can be used. Note that the Combinators.m package is not preloaded either, because it shadows system symbols.

In the example *Mathematica* sessions, we shall generally no longer show the command to read in the package that is the topic of the example. This command of the form <<MathProg`*Package*` or Needs["MathProg`*Package*`"] is assumed at the beginning of every session that uses functions from the package in question. *Mathematica* knows how to convert the context name into a file name that is valid on a given computer system. This is especially important on old-fashioned systems that do not support longer file names. If a program is a proper package, the latter form is preferred. If you read the init.m package in the way explained earlier, the Needs[] commands are unnecessary anyway.

Please observe that all examples have been computed in a fresh *Mathematica* session. While the package mechanism minimizes the chance of a conflict between different functions, it is an all too common error to forget about earlier global definitions.

> Some of the programs are also available through *MathSource* or have been distributed otherwise. Their copyright notice allows a wider distribution. The rest of the programs have not been made available through other channels and their distribution is restricted. Please consult the copyright notice in the files for more information. As owner of this book you are in any case allowed to use them and make any necessary archival copies. Smaller program examples bear no explicit copyright notice and are not restricted in distribution.

About This Book

Notation and Terminology

Mathematica input and output is typeset in typewriter-like style: `Expand[(x+y)^9]`. Genuine dialogue with *Mathematica* is set in two columns. The left column contains explanations and the right column contains the input and output, including graphics. Note that we do not take advantage of the new typesetting capabilities of the Version 3 frontend. They are not directly relevant to programming.

Parts of *Mathematica* input that are not literal, but denote (meta-)variables, are typeset in italics: `f[`*var_*`] := `*body*. *Functions* or *commands* are referred to by their name followed by an empty argument list, for example, `Expand[]`. Listings of programs are delimited with horizontal lines and usually have captions beneath them. Figures, listings, and tables are numbered in the form *c.s–n*, where *c* is the chapter number, *s* is the section number, and *n* is a consecutive number within each section. Most listings are printed near the place where they are discussed. Some of the longer ones, however, have been put into a section of their own at the end of the respective chapter.

Names of files are set in Helvetica typeface: ParametricPlot3D.m. Often, we give package names in a system-independent way in the form of a context name, for example, Graphics`Polyhedra`. This form of the name can be used to read the package in *Mathematica* with `<< Graphics`Polyhedra`` or `Needs["Graphics`Polyhedra`"]`. Notebooks have an extension of .nb.

The Mathematica Programmer CD-ROM

All programs discussed are included in the CD-ROM that comes with this book. It is a hybrid disk that contains partitions for Macintosh, MS-DOS/Windows, and UNIX. The disk contains all programs from both volumes of this book. Besides the programs, the CD-ROM contains additional information from the two volumes that is best accessed electronically. To make navigation around the information on the disk easier, I designed a hypertext access system, using the technique that is employed so successfully on the World Wide Web (WWW). Alternatively you can read the README.nb notebook using *Mathematica* or *MathReader*. It contains hyperlinks to all other notebooks.

The main entry point is the *table of contents*, accessible from the *top* or *index document*, called index.htm. The table of contents contains an anchor (link) for each chapter and other parts of the books. There is also some information not directly related to a particular chapter. It can be accessed from the top or index document.

All chapter contents pages contain the following information:

The chapter opener picture This image is taken from the title page of the chapter.

Overview This is the abstract of the chapter. It is taken from the text on the page following the chapter title.

Programs An annotated list of all programs mentioned in the chapter. The programs are anchors that let you read the program text.

Notebooks An annotated list of all notebooks relating to the chapter. Often, there is just one notebook that contains many of the calculations reproduced in the chapter. The names are anchors that let you start up *Mathematica* and load the notebook. You can then review (and modify!) the calculations.

Color Plates An annotated list (with thumbnail images) of all color images pertaining to this chapter. The anchors let you view the pictures and their captions at a higher resolution.

Additional Material This section may contain additional material not present in the book.

Resources on the Internet A list of anchors pointing to resources on the Internet that are related to the topic of the chapter. Among other things, the links give you access to the programs used for the graphic examples in the second part of the book. All of these programs are freely available and can be downloaded with the help of your WWW browser.

If you have a WWW browser, such as NETSCAPE, INTERNET EXPLORER, or MOSAIC, already installed on your computer, you should be able to load the CD-ROM and start reading the documents. The entry point is the document index.htm. If you do not have a browser, you need to install one first. The README document contains information on how to obtain a browser over the Internet. Some of them are freely available. Alternatively you can use *Mathematica* or *MathReader* as a browser by opening the README.nb notebook. It contains hyperlinks to most other information on the CD-ROM. *MathReader* is freely available from Wolfram Research; see the README document for more information.

The Mathematica Programmer WWW Archive

If you have Internet access, you can take advantage of additional information from *The Mathematica Programmer WWW Archive*. The archive is located somewhere on the Internet. It contains information that supersedes or augments the current information on the CD-ROM. Because the CD-ROM is a read-only device, I chose this approach to have a means of updating information. The archive pages will be updated from time to time. I welcome also your feedback. If you find errors or have other comments to make, please use the tools provided on the disk to let me know. The information in the archive is accessed through the html pages on the CD-ROM.

The archive contains one update page for each page on the CD-ROM. At the bottom of each page is an anchor icon that points to the corresponding update page. You can click the icon if you want to find out whether there is any new information for the page you are viewing.

Colophon

The manuscript was written in LaTeX [34] (with many extensions). It contains only the input of the numerous computation examples. The results were computed by *Mathematica* and then automatically inserted into the input file for TeX. The bibliography was produced with BibTeX [45] and the index was sorted with the help of makeindex [35]. Figures not produced with *Mathematica* were generated with FrameMaker on a NeXT computer and converted into PostScript. The output of TeX was then combined with all PostScript graphics, using dvips[51] and output on a phototypesetter.

The color images were rendered using a number of techniques explained in Chapters 11 and 12. The files were converted to TIFF for inclusion in the color insert.

Part 1

Paradigms of Programming

Chapter One

Introduction

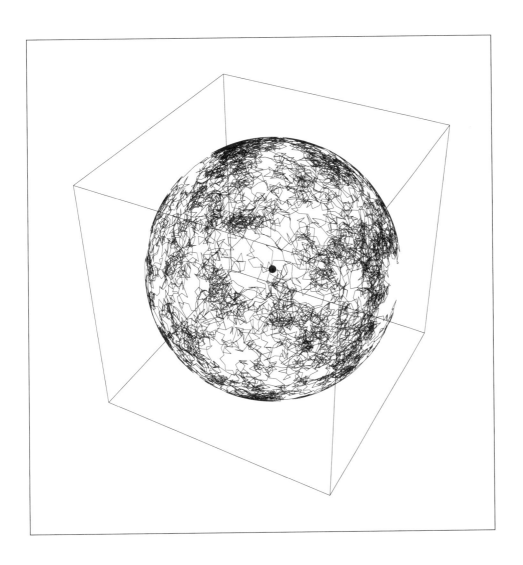

Mathematica includes a rich and powerful programming language. It combines the procedural, functional, and rule-based programming styles in a single coherent system.

Section 1 gives an overview over the language and discusses parameter-passing mechanisms. Section 2 discusses pattern matching and term rewriting. In Section 3 we compare the different programming styles in the way they present themselves in *Mathematica*. Section 4 discusses software engineering issues, such as modularization and information hiding.

About the illustration overleaf:

A spherical random walk. At each step, a random direction between 0 and 2π is chosen. This picture shows 10,000 steps. The code of the command SphereWalk[] is in the package Sphere-Walk.m. Methods to construct random walks are discussed in this chapter. A stereo pair of a random walk is reproduced in Section 11.5.

The code for this picture is in BookPictures.m.

1.1 *Mathematica*'s Programming Language

Mathematica has been available since 1988 and is being used more and more in teaching, research, and in industry. A by-product of the symbolic computation system is a programming language that differs from traditional languages in many important ways. It combines ideas from many different sources.

- Functional programming is well suited to express mathematical formulae and algorithms.
- Term rewriting allows for the expression of mathematical rules in a simple way.
- Powerful structural operations make the manipulation of composite data easy.
- Modularization is an important tool for organizing larger programs.
- Object-oriented elements make code development easier.
- Traditional procedural programming provides an easy entry point for those who are used to this programming style.

All of these ideas taken together lead to a large language with many built-in functions. Nevertheless, it has a consistent and uniform style. The reason is that is uses one underlying mechanism, term rewriting, on which all other programming constructs are based.

The language is interactive and, therefore, easy to use. It is not necessary to compile functions or to embed them into a main program to use them.

1.1.1 A Uniform Paradigm

Strictly speaking, there are no procedures, functions, or subroutines in *Mathematica*. Any definition of the form

$$f[args] := body$$

is a *rewrite rule*. Whenever the evaluator sees an expression that matches the left side (f[*args*]), it replaces the expression by the right side (*body*) with the values of the pattern variables substituted. This corresponds closely to a procedure call of a traditional language, a similarity that is intended. Definitions can, therefore, be used in place of traditional procedures, functions, and rewrite rules.

A procedure body typically declares some local variables and consists of a sequence of assignment statements and control structures.

```
SplitLine[vl_] :=
      Module[{vll, pos, linelist = {}, low, high},
             vll = If[NumberQ[#], #, Indeterminate]& /@ vl;
             pos = Flatten[ Position[vll, Indeterminate] ];
             pos = Union[ pos, {0, Length[vll]+1} ];
```

```
        Do[ low = pos[[i]]+1;
            high = pos[[i+1]]-1;
            If[ low < high,
                AppendTo[linelist, Take[vll, {low, high}]] ],
            {i, 1, Length[pos]-1}];
        linelist
    ]
```

A typical procedure

A function contains nested calls of other functions and returns a result.

```
RandomPoly[x_, n_] :=
        Sum[ Random[Integer, {-10, 10}] x^i, {i, 0, n} ]
```

A typical function

A rewrite rule transforms an expression into another one.

```
log[a_ b_] := log[a] + log[b]
```

A typical rewrite rule

1.1.2 Parameters of Procedures

An important difference from traditional procedural languages is the treatment of *formal parameters* of functions or procedures. Superficially, a definition in *Mathematica*, such as

$$f[x\_, y\_] := body$$

looks (intentionally) similar to a function declaration. The pattern variables in such a definition are, however, not local variables in the procedure (as they are in PASCAL or C)! The values of the arguments are *inserted* for every occurrence of the pattern variables in the body of the definition. This happens with substitution. The function call f[a, b] is evaluated essentially as

$$\texttt{ReleaseHold[Hold[}body\texttt{] /. \{x :> }a\texttt{, y :> }b\texttt{\}]}.$$

It is an instructive exercise to emulate the two standard parameter-passing mechanisms of procedural languages: *call by value* and *call by reference*.

Parameter Passing by Value

It is easy to emulate C-style procedure parameters:

1.1 Mathematica's Programming Language

```
f[x0_, y0_, z0_] :=
        Module[ {x=x0, y=y0, z=z0},
                    ⋮
         ]
```
C-style procedure parameters

The idea is to use initialized local variables. The variables x, y, and z are then proper local variables. This code corresponds to the following C declaration:

```
f(x, y, z)
{
     ⋮
}
```
Function parameters in C

This function finds the fixed point of $x \mapsto 1 + \frac{1}{x}$ with starting value x_0.

```
In[1]:= fixedpoint[x0_] :=
            Module[{x = x0},
                While[ x != 1 + 1/x, x = 1 + 1/x ];
                x
            ]
```

The fixed point is the *Golden Ratio*.

```
In[2]:= fixedpoint[1.2]
Out[2]= 1.61803
```

Exercise: What happens if the pattern variable is named x_ and Module[] is left out?

Parameter Passing by Reference

On the other hand, parameter passing by reference (var parameters in PASCAL, pointers in C, references in C++) is also possible. Here is the outline of a function in C++ that declares a reference parameter:

```
void f(int &x)
{
     ⋮
     x = val;
}
```
Parameter passing by reference in C++

In *Mathematica*, the same idea can be realized by a function with the attribute HoldFirst, HoldRest, or HoldAll.

```
SetAttributes[f, HoldAll]
f[xref_Symbol] :=
        Module[{locals...},
                ⋮
                xref = val (* assignment to global parameter *)
        ]
```

Parameter passing by reference

Because the attribute `HoldAll` prevents evaluation of the actual argument in a call of the function `f`, the unchanged arguments are inserted into the body of the function. Any computation done is therefore effectively performed on the argument itself. No other (local) variables are involved.

Attributes should be set before any rules are defined.	`In[3]:= SetAttributes[inc, HoldAll];`
This defines a function `inc` that increments its parameter.	`In[4]:= inc[nref_Symbol] := nref++`
Here is a global variable with value 5.	`In[5]:= var = 5;`
Because `inc` does not evaluate its argument, the operation is performed on the global variable `var`.	`In[6]:= inc[ var ]` `Out[6]= 5`
The value of `var` has indeed been changed.	`In[7]:= var` `Out[7]= 6`
It does not make sense to call `inc` with anything other than a symbol. The pattern does not match and nothing happens.	`In[8]:= inc[ 17 ]` `Out[8]= inc[17]`

Exercise: What happens if `inc` does not have the attribute `HoldAll`?

1.2 Pattern Matching and Term Rewriting

Pattern matching and term rewriting are the fundamental principle of *Mathematica*'s evaluator. All other programming constructs are implemented in terms of it. It is especially useful for implementing rules corresponding to transformations from a handbook of formulae. Equations in a handbook are usually meant to be used as rewrite rules, transforming the left side into the right side. By looking up an expression we essentially perform pattern matching in our head.

1.2 Pattern Matching and Term Rewriting

1.2.1 Example: Laplace Transforms

Definitions can be taken almost verbatim from a handbook of mathematics. Here is an excerpt from the standard package Calculus`LaplaceTransform`. It shows some simple rules for Laplace transforms of polynomials and exponential functions. The first three definitions express the fact that the Laplace transform is a linear transform.

```
LaplaceTransform[c_, t_, s_] := c/s /; FreeQ[c, t]
LaplaceTransform[c_ a_, t_, s_] := c LaplaceTransform[a, t, s] /; FreeQ[c, t]
LaplaceTransform[x_Plus, t_, s_] := LaplaceTransform[#, t, s]& /@ x
LaplaceTransform[t_^n_., t_, s_] := n!/s^(n+1) /; (FreeQ[n, t] && n > 0)
LaplaceTransform[a_ t_^n_., t_, s_] :=
      (-1)^n D[LaplaceTransform[a, t, s], {s, n}] /; (FreeQ[n, t] && n > 0)
LaplaceTransform[a_. Exp[b_. + c_. t_], t_, s_] :=
      LaplaceTransform[a Exp[b], t, s-c] /; FreeQ[{b, c}, t]
```

Part of LaplaceTransform.m

Packages can be referred to by their context name. This is a machine-independent specification.

```
In[1]:= Needs["Calculus`LaplaceTransform`"]
```

Here is the transform of $e^{\omega t + \varphi} t^2$.

```
In[2]:= LaplaceTransform[ Exp[omega t + phi] t^2, t, s]
```

$$\text{Out[2]}= \frac{2\,E^{phi}}{(-omega + s)^3}$$

1.2.2 Mathematical Programming

Mathematical concepts can easily be translated into programs. For example, the properties of the greatest common divisor (gcd) of two numbers

$$\begin{aligned} \gcd(a, b) &= \gcd(b, a \bmod b) \\ \gcd(a, 0) &= a \end{aligned} \qquad (1.2\text{--}1)$$

which lead to Euclid's algorithm, can be turned into a working program almost verbatim:

```
gcd[a_, b_] := gcd[b, Mod[a, b]]
gcd[a_, 0]  := a
```

The definitions can be used immediately.

```
In[1]:= gcd[1999, 2999]
Out[1]= 1
```

The built-in tracing and debugging facilities make it easy to study how the program works. Here we see the sequence of calls of gcd.

```
In[2]:= Trace[ gcd[5, 8], gcd[__Integer] ] // TableForm
Out[2]//TableForm= gcd[5, 8]
                   gcd[8, 5]
                   gcd[5, 3]
                   gcd[3, 2]
                   gcd[2, 1]
                   gcd[1, 0]
```

Term rewriting and its connections with programming paradigms are treated in more detail in [39, 41].

1.3 Programming Styles

This section shows examples of some commonly identified programming styles.

1.3.1 Functional Programming

Given the problem of adding the square roots of the first 500 integers, the solution in most programming languages is to use an auxiliary variable that is incremented by the square root of a loop index iterating from 1 to 500.

```
sum = 0.0;
Do[ sum = sum + N[Sqrt[i]], {i, 1, 500} ];
sum
```

In symbolic computation systems, this loop reduces to a single statement

$$\text{Sum[N[Sqrt[i]], \{i, 1, 500\}]}$$

which corresponds directly to the mathematical formula

$$\sum_{i=1}^{500} \sqrt{i}. \tag{1.3-1}$$

Mathematica does not force you to think about *how* to implement a summation, instead it lets you focus on the *concept* itself.

Functional programming is the topic of Section 3.3.

1.3 Programming Styles

1.3.2 Structural Operations

Powerful operations allow us to treat large structured objects (vectors, matrices, nested lists of things) as single objects. This avoids most of the loops found in traditional programming languages. (It is interesting to note that FORTRAN, the language used most often for numerical computation, lacks operations on vectors and matrices as a whole. One always has to write a loop.) Many combinatorial problems can be solved with structural operations.

Distribution

Distribute[$\{l_1, \ldots, l_n\}$, List] can be used to generate all ways of choosing one element each from the lists $l_1, \ldots, l_n$.

There are six ways to choose such elements.

```
In[1]:= Distribute[{ {a1, a2}, {b1, b2, b3} }, List]
Out[1]= {{a1, b1}, {a1, b2}, {a1, b3}, {a2, b1}, {a2, b2},
         {a2, b3}}
```

If one of the l_i is not a list, it is treated as a list with one element. Here is the list of all possibilities of choosing one of a1, a2, and a3, one b and one of c1, c2.

```
In[2]:= Distribute[{{a1, a2, a3}, b, {c1, c2}}, List]
Out[2]= {{a1, b, c1}, {a1, b, c2}, {a2, b, c1},
         {a2, b, c2}, {a3, b, c1}, {a3, b, c2}}
```

Here is a small application: a random walk on a grid. To generate the path of a particle, we determine the directions to all neighboring points, define a random variable whose values are one of these directions, then generate a list of steps, and, finally, compute and illustrate the path.

This is the list of all grid points next to the origin (in two dimensions).

```
In[3]:= Distribute[{{-1,0,1}, {-1,0,1}}, List]
Out[3]= {{-1, -1}, {-1, 0}, {-1, 1}, {0, -1}, {0, 0},
         {0, 1}, {1, -1}, {1, 0}, {1, 1}}
```

We get rid of the origin itself; all neighboring points remain.

```
In[4]:= neighbors = Complement[%, {{0, 0}}]
Out[4]= {{-1, -1}, {-1, 0}, {-1, 1}, {0, -1}, {0, 1},
         {1, -1}, {1, 0}, {1, 1}}
```

This definition chooses a random direction each time the symbol randDir is evaluated. Note the use of delayed assignment :=.

```
In[5]:= randDir :=
        neighbors[[Random[Integer, {1, Length[neighbors]}]]]
```

Here are 1000 random steps in one of the 8 directions.

```
In[6]:= (steps = Table[ randDir, {1000} ]) // Short
Out[6]//Short=
   {{1, 0}, {0, -1}, {0, 1}, {1, 1}, {-1, -1}, {0, -1},
    {-1, 0}, <<990>>, {-1, -1}, {1, -1}, {1, -1}}
```

Their cumulative sums give us the locations visited in turn.	`In[7]:= (locs = FoldList[Plus, {0,0}, steps]) // Short`
	`Out[7]//Short=`
	`  {{0, 0}, {1, 0}, {1, -1}, {1, 0}, {2, 1}, {1, 0}, {1, -1},`
	`   <<991>>, {34, -10}, {35, -11}, {36, -12}}`

With no effort at all, we get an illustration of this random walk.	`In[8]:= Show[ Graphics[{Line[locs]}],` `            PlotRange -> All, AspectRatio -> Automatic ];`

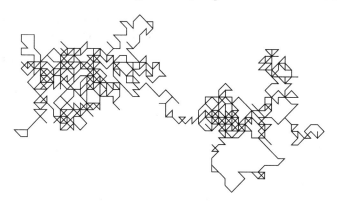

In Section 10.2.2, we shall use `Distribute[]` to enumerate all stellations of the icosahedron.

Transposition

Transposition is useful not only for ordinary matrices, but also to rearrange data in various ways. Example: How do we find the permutation that sorts a list into standard order?

Here is a sample list.	`In[9]:= l = {1.2, 5.4, -2.1, 33, 18.5};`
We attach the index of each number in a list of pairs.	`In[10]:= Transpose[{l, Range[Length[l]]}]` `Out[10]= {{1.2, 1}, {5.4, 2}, {-2.1, 3}, {33, 4}, {18.5, 5}}`
Now we can sort it.	`In[11]:= Sort[ % ]` `Out[11]= {{-2.1, 3}, {1.2, 1}, {5.4, 2}, {18.5, 5}, {33, 4}}`
We keep only the indices; they describe the original position of the numbers, that is, the permutation.	`In[12]:= tr = Transpose[%][[2]]` `Out[12]= {3, 1, 2, 5, 4}`
The permutation can be used to rearrange other lists of dependent information into the same order.	`In[13]:= {a, b, c, d, e}[[tr]]` `Out[13]= {c, a, b, e, d}`

We shall use this idea in Section 9.3 to sort the face types of a uniform polyhedron.

1.3 Programming Styles

Positional Operations

Map performs the same function on all elements of a structure. Sometimes, the function to perform depends on the position of the element in the structure. MapIndexed[f, *structure*] calls f on every element of *structure*, just like Map[], but it passes the index of the element as a second argument to f.

Here we see what it does. The index is wrapped into list braces.	In[14]:= `MapIndexed[ f, {e1, e2, e3} ]` Out[14]= `{f[e1, {1}], f[e2, {2}], f[e3, {3}]}`

Here, g is applied to all elements of a matrix. The indices are lists of length 2.	In[15]:= `MapIndexed[ g, {{a, b}, {c, d}, {e, f}}, {2} ]`
	Out[15]= `{{g[a, {1, 1}], g[b, {1, 2}]},`
	`         {g[c, {2, 1}], g[d, {2, 2}]},`
	`         {g[e, {3, 1}], g[f, {3, 2}]}}`

If the extra braces around the index are not needed, suitable functions can be declared with the pattern `f[e, {i_}] := ...` or `g[e, {i_, j_}] := ...`. The indices are then available directly in the right sides of the definitions.

As an application, let us look at a class of matrix predicates for properties of a matrix $m = (m)_{ij}$ that depend only on the elements and their positions. The condition is satisfied if it is true for all elements. The matrix checker looks like this.

```
CheckMatrix[m_?MatrixQ, pred_] := Apply[ And, MapIndexed[pred, m, {2}], {0,1} ]
```

It takes a suitable predicate, calls it on each element, passing the element's index as a second argument, and forms the logical AND of all results.

This symbolic example shows how it works.	In[16]:= `CheckMatrix[{{a, b}, {c, d}}, p]` Out[16]= `p[a, {1, 1}] && p[b, {1, 2}] && p[c, {2, 1}] &&` `         p[d, {2, 2}]`

A number of elementary matrix properties can be checked in this way. Here are three examples.

This predicate checks that off-diagonal elements are zero. Matrices that satisfy it are diagonal.	In[17]:= `diagQ[e_, {i_, j_}] := i==j || e == 0;\` `         diagonalQ[m_] := CheckMatrix[m, diagQ]`
This matrix is diagonal.	In[18]:= `diagonalQ[ {{1, 0},{0, 2}} ]` Out[18]= `True`
Symbolic examples are simplified nicely. The matrix is diagonal if and only if the two off-diagonal elements are zero.	In[19]:= `diagonalQ[ {{a, b}, {c, d}} ]` Out[19]= `b == 0 && c == 0`

Matrices where all elements below the diagonal are zero are upper triangular matrices.	`In[20]:= utQ[e_, {i_, j_}] := i<=j || e == 0;\` `        upperTriangularQ[m_] := CheckMatrix[m, utQ]`
Tridiagonal matrices have nonzero elements only along the diagonal and adjacent to it.	`In[21]:= triDQ[e_, {i_, j_}] := Abs[i-j] <= 1 || e == 0;\` `        triDiagonalQ[m_] := CheckMatrix[m, triDQ]`

Not all properties can be checked in this way. To test whether a matrix is symmetric, for example, the elements must be compared. The property cannot be decided by looking at single elements at a time.

Of course, to test whether a matrix is symmetric, it is much more elegant and efficient to use a structural operation than it is to run a double loop over all elements.	`In[22]:= symmetricQ[m_?MatrixQ] := m == Transpose[m]`
This form does not apply any logical simplifications, however.	`In[23]:= symmetricQ[ {{a, b}, {c, d}} ]` `Out[23]= {{a, b}, {c, d}} == {{a, c}, {b, d}}`
Here is the desired condition for the matrix $\begin{pmatrix} a & b \\ c & d \end{pmatrix}$ to be symmetric.	`In[24]:= LogicalExpand[ % ]` `Out[24]= c == b`

1.3.3 Lazy Evaluation

With lazy evaluation, arguments of functions are only evaluated when they are needed, not when the function is called. It can be implemented with the attributes `HoldFirst` and `HoldRest`. Among other things this idea allows us to implement *streams* (infinite lists); see Chapter 6 of the first volume of this book. Here is another random walk, generated with a stream (compare with the example on page 11).

We read the package that defines streams.	`In[25]:= Needs["MathProg`Streams`"]`
This function generates an unbounded random walk. Each next location is equal to the previous one plus a random step.	`In[26]:= randomWalk[from_] :=` `            MakeStream[ from, randomWalk[from+randDir] ]`
Here is a random walk, starting at the origin.	`In[27]:= walk = randomWalk[{0,0}]` `Out[27]= {{0, 0}, randomWalk[<<1>>] ...}`

1.3 Programming Styles

We can plot any initial segment in the same way we did earlier.

```
In[28]:= Show[ Graphics[{Line[Take[walk, 1000]]}],
                PlotRange -> All, AspectRatio -> Automatic ];
```

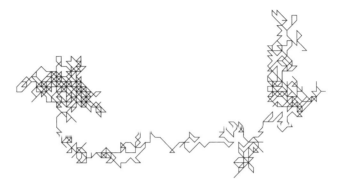

1.3.4 Logic Programming

Logic programming or *declarative programming* tries to write declarations that express certain properties of the desired result without specifying the flow of control. Pattern matching and backtracking are then used to solve an instance of the problem.

As an example, here is a program for reversing a list in typical PROLOG style.

```
reverse[l_] := rev[ l, {} ]
rev[ {}, r_ ] := r
rev[ {h_, t___}, {r___} ] := rev[ {t}, {h, r} ]
```

The auxiliary function rev[*left*, *right*] operates on two stacks (lists). If the left list is empty, we are done; otherwise, we remove the first element of the left list and prepend it to the right list. Removing the first element from the left list is done implicitly, by using a pattern {h_, t___}.

A trace shows how this code works. It uses the method we humans would use to reverse the order of a deck of cards: taking one after another and placing it onto a second pile.

```
In[1]:= TracePrint[ reverse[{a, b, c, d, e}], rev[_,_] ]
    rev[{a, b, c, d, e}, {}]
    rev[{b, c, d, e}, {a}]
    rev[{c, d, e}, {b, a}]
    rev[{d, e}, {c, b, a}]
    rev[{e}, {d, c, b, a}]
    rev[{}, {e, d, c, b, a}]
Out[1]= {e, d, c, b, a}
```

Backtracking can be implemented with side conditions. The pattern matcher generates all possible cases. The side condition can then be used to commit a certain case. Lists can be sorted in this way.

An *inversion* in a list is a pair of adjacent elements such that the first one is larger than the second one. A sorted list is characterized by not having any inversions of adjacent elements. To sort a list, we simply transpose any inversions found. If none are left, the list is sorted. No particular order of doing this needs to be specified.

```
sort[ {alpha___, x_, y_, omega___} ] /; x > y := sort[ {alpha, y, x, omega} ]
sort[ l_ ] := l
```

The pattern matcher generates all possible pairs of adjacent x and y. Whenever they are out of order, they are reversed.

```
In[2]:= sort[ {5, 1, 3, 2} ]
Out[2]= {1, 2, 3, 5}
```

ReplaceList[*expr, rule*] gives a list of the results of all possible ways to match the left side of *rule* to *expr*. It is a useful debugging tool for complicated rules. Here we use it to pick out the inversions in our list.

```
In[3]:= ReplaceList[ {5, 1, 3, 2},
            {alpha___, x_, y_, omega___} /; x > y :> {x, y}
        ]
Out[3]= {{5, 1}, {3, 2}}
```

We shall develop an interpreter for the language PROLOG in Chapter 2.

1.3.5 Abstract Data Types

Abstract data types are both a theoretically well-defined concept and a useful tool for program development. Following the principles of abstract data type design, you achieve a clear separation of *specification* and *implementation*.

Abstract data types are defined in terms of type names, function names, and equations. These can be realized in *Mathematica* easily. The equations become rewrite rules. The interactive nature of *Mathematica* makes it well suited for rapid prototyping and testing of designs. We discussed abstract data types in Chapter 2 of the first volume of this book.

Abstract data types are designed according to these principles:

- Define the type names to be used.

- Define the constants to be used.

- Define constructors, selectors, and predicates.

- Express the relations among the terms as equations.

1.3 Programming Styles

The implementation proceeds according to these steps:

- Choose a representation for elements of the types defined.
  Usually you can use a normal expression having the type name as head. This corresponds roughly to a record in many programming languages.
- Derive rules for normal forms for the data elements.
  This ensures that elements are stored in a unique way. The rules are derived from the equations and can normally be put into the constructors.
- Define constructors, selectors, and predicates.
  These are the only operations that are allowed access to the details of the internal representations.
- Define the other operations.
  An operation for which exactly one equation exists can usually be turned into a simple rewrite rule. The variables in the left side of the rules are of the form n_type, restricting the arguments of the functions implemented to the correct types. Use selectors only to access parts of the data elements on the right side of the rules.
- Use overloading where appropriate.
  Overloading built-in functions is most easily done by defining upvalues, rules of the form

$$g/\colon f[\mathtt{n\_}g,\ldots] := \ldots.$$

- Choose a suitable output representation of data elements.
  If you do this, you can consider declaring the type names themselves in the private implementation part of your package. Users are then prevented from accessing them directly.

1.3.6 Object-Oriented Programming

Object-oriented programming is a programming style that is becoming more and more popular. It promises code reuse and easier maintenance of larger projects than is possible with traditional procedural languages. Its use of methods and message passing instead of procedure calls shifts the programmer's view toward close integration of data and operations. We treated object-oriented programming in Chapter 4 of the first volume of this book.

Objects and Methods

The combination of data and operations at the level of single objects is the first important idea behind object-oriented programming. An *object* is, therefore, a collection of data elements together with the functions operating on them. These functions are called *methods*. A general mechanism, *message passing*, is used to invoke a method. We say that a message is passed to an object to perform a method. We shall represent message passing as an ordinary function call, that is, as

$$message[object, arguments...]$$

The object to which the message is sent is called the *receiver* of the message.

Classes

Methods are normally not defined for each object separately, but only for classes of objects. A class is characterized by the fact that all objects belonging to this class know the same methods, that is, they show they same behavior.

In addition to the methods, objects contain also data elements. The data elements of an object are called *instance variables*. Such an instance variable is like a component in a composite data type. All objects in a class have the same set of variables, but each object has its own private values for these variables. An object is also called an *instance* of its class.

Inheritance

The second important concept in object-oriented programming is *inheritance*. Often, related data types have common characteristics.

We can take advantage of these common aspects and program some of the methods in a way that does not depend on which of the similar classes an object belongs to. The program works in the same way for objects from any of the related classes. The common characteristics can, therefore, be extracted and can be put into a new class. The other classes are then made subclasses of the new class. They inherit instance variables and methods from their superclass, and need to define only those variables and methods that are special to them. This code sharing in a superclass makes programs easier to maintain.

1.4 Program Organization

Good software design principles dictate that interfaces be specified clearly and that implementation details be kept hidden and local to the program unit that contains them. There are two levels at which these principles manifest themselves in *Mathematica*: the package or program unit, and the individual procedure or function.

1.4 Program Organization

1.4.1 Modularization

Modern programming languages provide features for organizing large programming projects. Most important are *modularization* (or encapsulation) and *information hiding*. A *Mathematica* program that is organized according to these principles is called a *package*. A package consists of two parts: an *interface definition* and an *implementation part*. The interface describes the aspects that a client needs to know. The implementation provides that functionality but it is hidden from users of the package. Here is how such a package looks like in *Mathematica* (this example is taken from [36]):

```
BeginPackage["Reshape`"]
Reshape::usage = "Reshape[list, dims] rearranges list
    as a nested structure with dimensions dims."
Begin["`Private`"]
reshape[list_, {n_}] /; Length[list] == n := list
reshape[list_, {head__, n_}] := reshape[ Partition[list, n], {head} ]
Reshape[list_, dims_] := reshape[ Flatten[list], dims ]
End[]
Protect[ Reshape ]
EndPackage[]
```

The part between the initial `BeginPackage[]` and `Begin["`Private`"]` is the interface. It declares all functions exported from this package (there is only one: `Reshape`) and documents them. The argument of `BeginPackage[]` is the context in which all exported symbols will be defined. It is usually also the name of the file (with an extension .m). The package is used by *importing* it, in the form `Needs["Reshape`"]`.

The part between `Begin[]` and `End[]` is the implementation. The initial context mark ` in `` `Private` `` makes this context a subcontext of the current context, which is the package context `Reshape``. The auxiliary function `reshape` is not exported. It will not be available to users of this package. The command `Protect[Reshape]` prevents any unintentional modification of the exported function `Reshape` by users of the package.

The design of packages and the additional possibilities that *Mathematica*'s contexts offer are explained in [40].

1.4.2 Scoping

On the level of a single procedure or function the main concern is to make its operation independent of the context in which it may be invoked. The main tool is the declaration of local variables. Any auxiliary variables should be declared in a module of the form

```
f[x_, y_] :=
    Module[{z},
        ⋮
    ]
```

The variable z is local to the function f and does not interfere with global variables of the same name. Formal parameters of functions and pattern variables are also treated as local variables, as these experiments show.

We use a global variable named x.	`In[1]:= x = 42;`
Here is a simple pure function, using x as the name of its formal parameter.	`In[2]:= f = Function[x, x^2];`
Application and evaluation of the function is not disturbed by the global variable x.	`In[3]:= f[3]` `Out[3]= 9`
Here we define a function g, using x as a pattern variable.	`In[4]:= g[x_] := x^3`
This use of x does not conflict with the global x either.	`In[5]:= g[3]` `Out[5]= 27`
There is an important exception: variables in immediate values are not treated as local before the right side of the definition is evaluated.	`In[6]:= h[x_] = x^4` `Out[6]= 3111696`
As a result, the function h is constant, that is, independent of its argument.	`In[7]:= h[1]` `Out[7]= 3111696`

 The CD-ROM contains the package SphereWalk.m.

Chapter Two

Logic Programming

This chapter is about logic programming. The first section gives an introduction. Because logic programming is not built into *Mathematica*, our first task is to develop an interpreter for a logic language. We do so in Section 2. This query evaluator is an interpreter for a subset of the programming language PROLOG. A prerequisite for it is *unification,* which is a generalization of the pattern matching that underlies *Mathematica*'s own evaluator.

In the remainder of the chapter, we use the query evaluator to give examples of typical logic programming applications. Among the examples considered are implementations of PROLOG-style lists (Section 3), nondeterministic automata, backtracking, and exhaustive search (Section 4), as well as theorem proving and deductive databases (Section 5).

About the illustration overleaf:
A variant of the Sierpinski carpet as the invariant set of an iterated function system consisting of five affine contractions. Shown are seven iterations leading to 78,125 points. Iterated function systems are treated in [40].

The code for this picture is in BookPictures.m.

2.1 The Ingredients of Logic Programs

In functional programming, a program consists of a number of function definitions. The body of a function applies other functions to its arguments. In a procedural program, a sequence of instructions, including conditional statements, loops, and assignments, performs a computation step by step. In a *logic* or *declarative* program, a number of *inference rules* is given. A query is then formulated and the evaluator tries to prove it, given the knowledge of the rules in its database. No procedural interpretation is intended in a pure logic language. The rules simply state logic properties about certain predicates. Because any evaluator for such a language defines implicitly procedural or operational semantics, we will encounter differences between mathematical logic and a logic language. Nevertheless, we will clearly see the distinction between logic programming and traditional procedural programming. The examples in Sections 2.3–2.5 show the types of tasks for which a logic language is much better suited than any other kind of language.

Listing 2.1–1 shows a simple logic program that shall serve as our main example for showing how the query evaluator works and what can be done with logic programming.

```
Assert[ arc[a, b] ]
Assert[ arc[a, c] ]
Assert[ arc[c, d] ]

Assert[ path[x_, x_] ]
Assert[ path[x_, z_], arc[x_, y_], path[y_, z_] ]
```

Listing 2.1–1: DAG.m: Logic rules for a DAG

The program describes a small *directed acyclic graph* (DAG) depicted in Figure 2.1–1. In such a graph, edges (called *arcs*) are directed, that is, they can be traversed only in one direction (like one-way streets). There is an arc from vertex a to b, but not from b to a. A *path* is any sequence of consecutive arcs. Because there is an arc from a to c and one from c to d, there is a path from a to d. There is also a (trivial) path from a to a.

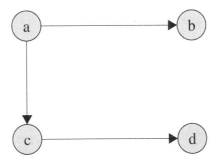

Figure 2.1–1: A small DAG

Note that we translated the syntax used commonly in PROLOG into ordinary *Mathematica* syntax. The logic rules are given as arguments to the function `Assert[]`. In PROLOG, the example would look as shown in Listing 2.1–2.

```
arc(a, b).
arc(a, c).
arc(c, d).
path(X, X).
path(X, Z) :- arc(X, Y), path(Y, Z).
```

Listing 2.1–2: The program from Listing 2.1–1 in PROLOG

All logic statements are terminated by a period. The left side of the rule (the conclusion) is separated from the right side (the preconditions) by the operator `:-`. Preconditions are separated by commas.

The first three rules have no right side. Such rules are called *facts*. They simply state properties that are unconditionally true and they correspond to axioms in logic. The fourth rule is also a fact, but it contains a *variable*, denoted by the familiar underscore in *Mathematica* and by *capitalization* in PROLOG. This rule simply means that for any value of x whatsoever there is a path from x to x. The last rule says that there is a path from x to z *provided* there is an arc from x to some y *and* there is a path from y to z. This rule has two clauses on the right side, implying a conjunction: *both* must be satisfied for the conclusion to be valid. Note that we must use the underscore also on the right side of the rules, which is different from ordinary rules in *Mathematica*. I can predict from experience that you will forget this underscore many times before you get used to it. Because both the query and the rule may contain variables, there is no way around this.

I have just explained the meaning of the short example program. What can we do with it? If we assert these logic rules in a logic programming language we can ask questions about arcs and paths. This concept is very powerful, as we shall demonstrate with a few examples.

First, we have to read in the logic program evaluator.

`In[1]:= Needs["MathProg`LogicProgramming`"]`

We also read the program from Listing 2.1–1.

`In[2]:= << MathProg/DAG.m`

The two rules for the predicate path have been defined. Note, that the two clauses in the second one have been expressed as an explicit conjunction (using *Mathematica*'s logic And function, written as `&&`).

```
In[3]:= LogicValues[ path ] // TableForm
Out[3]//TableForm=
 path[x_, x_]
 path[x_, z_] :- (arc[x_, y_] && path[y_, z_])
```

If we now ask whether there is an arc from a to c, the evaluator starts comparing our query `arc[a, c]` with the left sides of the logic values for `arc`. Because no variables are involved, the procedure is quite simple. The first fact does not compare, but the second one does.

2.1 The Ingredients of Logic Programs

The answer we get is simply "yes," meaning that the query can be fulfilled.	`In[4]:= Query[ arc[a, c] ]` `Out[4]= Yes`

Now let us ask "for which values of x is there an arc from a to x?" Our query is `arc[a, x_]`. It can be unified with the first two rules for `arc`, with the bindings x -> b and x -> d, respectively.

The answer of the evaluator is the first succeeding binding.	`In[5]:= Query[ arc[a, x_] ]` `Out[5]= {x -> b}`
We can ask the evaluator to *try again*! Now it finds the second successful binding.	`In[6]:= Again[]` `Out[6]= {x -> c}`
There are no more possibilities, so on the next try the evaluator says "no."	`In[7]:= Again[]` `Out[7]= No`

The command `QueryAll` prints all possible answers.	`In[8]:= QueryAll[ arc[a, x_] ]`
	`{x -> b}`
	`{x -> c}`

For a query with two variables, there are three possible unifications with all three rules.	`In[9]:= QueryAll[ arc[x_, y_] ]`
	`{x -> a, y -> b}`
	`{x -> a, y -> c}`
	`{x -> c, y -> d}`

Questions involving paths are more interesting because they involve a rule that has a right side. Let us ask "for which v is there a path from a to v?" We have to unify the query `path[a, v_]` with the left sides of the two rules for `path`. The first answer is immediate:

The query returns with the binding for the variables occurring in the query. With v set to a, the first rule for `path` is applicable.	`In[10]:= Query[ path[a, v_] ]` `Out[10]= {v -> a}`

If we try again, we have to turn to the second rule. To fulfill it, the queries in its right side have to be fulfilled first.

There is an arc from a to b and a path from b to b, so there is a path from a to b.	`In[11]:= Again[]` `Out[11]= {v -> b}`

A few additional possibilities can be obtained by trying some of the auxiliary queries again. Here are all solutions, that is, all points that can be reached from a.	`In[12]:= QueryAll[ path[a, v_] ]`
	`{v -> a}`
	`{v -> b}`
	`{v -> c}`
	`{v -> d}`

Let us look at some more examples of typical queries.

These are all points that can reach b.

```
In[13]:= QueryAll[ path[u_, b] ]
{u -> b}
{u -> a}
```

Queries can also contain several clauses. This query asks for all points reachable from a, excluding a itself.

```
In[14]:= QueryAll[ path[a, v_], v_ != a ]
{v -> b}
{v -> c}
{v -> d}
```

This query tests whether our graph is indeed acyclic. In an acyclic graph, there cannot be a path from u to v and back. Note that we must exclude the trivial case $u = v$.

```
In[15]:= Query[ path[u_, v_], u_ != v_, path[v_, u_] ]
Out[15]= No
```

We can also use anonymous pattern variables. This query could be phrased as "is there a path starting at a (leading somewhere)?" The anonymous variable is not shown in the result.

```
In[16]:= Query[ path[a, _] ]
Out[16]= Yes
```

2.1.1 Unification

The process of *unification* should be easy to understand for *Mathematica* users, because a weaker form of it—*pattern matching*—is the fundamental operating principle of *Mathematica*'s evaluator. In pattern matching, a *pattern* is compared with an expression. Pattern matching means finding replacements for the *pattern variables* occurring in the pattern that make the instantiated pattern identical to the expression. If this is possible, we say that the expression matches the pattern. Here is an example:

When *Mathematica* applies a rule, it tries to match expressions occurring in f[3, h[3]] with the pattern f[x_, h[x_]], which forms the left side of the rule. Here it succeeds with x -> 3 and then applies this substitution to the right side of the rule.

```
In[1]:= f[3, h[3]] /. f[x_, h[x_]] :> x
Out[1]= 3
```

This expression does not match and it is left alone.

```
In[2]:= f[3, h[4]] /. f[x_, h[x_]] :> x
Out[2]= f[3, h[4]]
```

The pattern matcher itself, which tries to find the variable bindings, is not available as a user-callable function. We provide a simple unifier in the package Unify.m. It takes two patterns as arguments and returns the list of bindings or $Failed, if a match cannot be found.

2.1 The Ingredients of Logic Programs

We read in the unifier package.
```
In[1]:= Needs["MathProg`Unify`"]
```

This is the matching that takes place in the previous example. If the pattern variable x is set to 3, the pattern becomes identical to the expression.
```
In[2]:= Unify[ f[x_, h[x_]], f[3, h[3]] ]
Out[2]= {x -> 3}
```

This expression does not match the pattern.
```
In[3]:= Unify[ f[x_, h[x_]], f[3, h[4]] ]
Out[3]= $Failed
```

But it matches this pattern with two pattern variables.
```
In[4]:= Unify[ f[x_, h[y_]], f[3, h[4]] ]
Out[4]= {x -> 3, y -> 4}
```

The function Unify[] is not as sophisticated as the built-in matcher. It understands only simple pattern variables in the form *name_* and does not take attributes into account (that is, it is not an associative/commutative pattern matcher).

In one important respect our function is more powerful than the built-in pattern matcher: it allows *unification*. Unification is two-way matching, where both arguments of the Unify function can contain pattern variables. It returns the most general variable substitution making the two patterns identical.

If we set x to a and y to b the two patterns become identical.
```
In[5]:= Unify[ f[x_, a], f[b, y_] ]
Out[5]= {x -> b, y -> a}
```

In this case, no substitution can be found because x would have to be equal to a and b simultaneously.
```
In[6]:= Unify[ f[x_, a], f[b, x_] ]
Out[6]= $Failed
```

Comparing the first arguments of f, we find that x -> g[3]. Comparing the second arguments with this binding in mind we find that g[y_] must unify with g[3], which gives the binding for y.
```
In[7]:= Unify[ f[x_, g[y_]], f[g[3], x_] ]
Out[7]= {x -> g[3], y -> 3}
```

With x -> g[y] the two patterns are identical, whatever value we choose for y. The result, therefore, does not contain a binding for y.
```
In[8]:= Unify[ f[x_], f[g[y_]] ]
Out[8]= {x -> g[y]}
```

Similarly here: If the two variables have identical values, the two patterns become identical, whatever that value is. Therefore, we replace one of the variables by the other one.
```
In[9]:= Unify[ g[x_, y_], g[y_, y_] ]
Out[9]= {y -> x}
```

Unification is a fairly simple process. Two identical expressions always unify without any bindings. A variable unifies with any expression by being bound to it. If the two expressions are composite, they must have the same head and the same number of elements; elements are then unified in turn.

Bindings obtained in this way are merged, and if a conflict is found, unification fails. The code is in the package Unify.m (not listed here).

Unification plays the same role in query evaluation as matching does in *Mathematica*: a list of rules is searched sequentially until one is found that unifies with the query. The unifying variable binding is then applied to the right side of the rule, and query evaluation continues.

2.1.2 Query Evaluation

A query establishes a *goal* to be satisfied. To do so, further subgoals may have to be satisfied. If a goal fails, the evaluator will try to resatisfy an earlier goal: it performs *backtracking*. If none of the subgoals can be redone, the original goal fails.

If we view the collection of rules for a predicate as a function, this function has *four* connections to its environment, unlike an ordinary function, which has just two: call and return. When we try a goal for the first time, we *call* the predicate. There are two possible outcomes: *success* and *failure*. In case of success we can later *redo* the predicate. This is often depicted as in Figure 2.1–2.

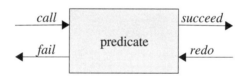

Figure 2.1–2: The procedure box model

For purposes of backtracking a conjunction of goals $goal_1$ && $goal_2$ && ... && $goal_n$ can be thought of as a connection of the corresponding procedure boxes, as shown in Figure 2.1–3. A succeeding call will invoke the next procedure until we succeed with the last one and thus with the complete goal. If one of the predicates fails, it invokes the redo entry of the preceding one. Control may pass back and forth many times until we either succeed with the last goal or fail the first one.

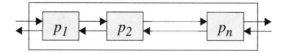

Figure 2.1–3: Backtracking a conjunction of goals

Let us return to the query path[a, v_] from the introduction and show in detail how the answers were obtained. We have to unify the query with the left sides of the two rules for path. The first answer is immediate.

2.1 The Ingredients of Logic Programs

The unification with the left side path[x_, x_] of the first rule succeeds with these bindings.	In[1]:= Unify[path[x_, x_], path[a, v_]] Out[1]= {x -> a, v -> a}
The query returns only the bindings for the variables occurring in the query itself.	In[2]:= Query[path[a, v_]] Out[2]= {v -> a}

If we try again, we have to turn to the second rule. Let us now show step-by-step what happens inside the evaluator.

The unification with the left side of the second rule succeeds with these bindings.	In[3]:= Unify[path[x_, z_], path[a, v_]] Out[3]= {x -> a, z -> v}

The values for x and z just found need to be inserted in the right side of the second rule. It becomes arc[a, y_] && path[y_, v_]. (Note that the value of z is again a variable.)

Now, we try to fulfill the first clause of the conjunction by a recursive call of the evaluator. We succeed with this binding for y.	In[4]:= Query[arc[a, y_]] Out[4]= {y -> b}

Now we apply this additional binding to the remaining clauses of the right side (there is only one more left) and get path[b, v_] as the remaining goal to satisfy.

Again a recursive call is used to fulfill this query. It succeeds by using the first rule for path. The result is the binding for our original query variable v.	In[5]:= Query[path[b, v_]] Out[5]= {v -> b}

To find additional solutions to our original query path[a, v_], we use *backtracking*. We turn to the last subgoal satisfied and try to redo it.

There is no other way to satisfy path[b, v_], that is, there is no arc starting at b.	In[6]:= Again[] Out[6]= No
Because the second clause could not be redone, we turn to the first one: arc[a, y_]. It can be redone, producing the next solution.	In[7]:= Again[] Out[7]= {y -> c}

We apply the new binding for y to the second clause and try to satisfy it (for the first time), producing our second solution to the original query.

```
In[8]:= Query[ path[c, v_] ]
Out[8]= {v -> c}
```

If we try our original query again, we backtrack and try to redo the last query. This time there is an additional solution. Backtracking therefore stops right here and we get the third solution.

```
In[9]:= Again[]
Out[9]= {v -> d}
```

2.2 A PROLOG Interpreter for *Mathematica*

Our query evaluator has become quite sophisticated. Nevertheless, it shows that a more-or-less complete PROLOG interpreter turns out to be a rather short program.

Internally, pattern variables like $x\_$ are represented as Var[x]. This representation protects them from *Mathematica*'s evaluator, for which pattern variables are also meaningful. The code for converting back and forth is in the auxiliary package Unify.m. The user-level unifier Unify[$pattern_1$, $pattern_2$] converts the pattern variables to internal form, calls the internal code in Unify0, and finally restores the pattern variables for output.

```
Unify[x_, y_] := Unify0[ x /. pattern2Var, y /. pattern2Var ] /. var2Symbol
```

The value of pattern2Var is a list of two rules for turning $x\_$ into Var[x] and $\_$ into Var[$, n], where n is an integer that is different for each anonymous variable encountered. The exact form of these rules is a case study in writing patterns for manipulating patterns. A pattern object can be protected from *Mathematica*'s pattern matcher by enclosing it in Verbatim[*pattern*]. However, if *some* part of the pattern should be treated as a pattern variable, you can enclose the *head* of the pattern (one of Blank, Pattern, Optional, etc.) in Verbatim[*head*]. The two rules show both cases.

```
anon=0 (* counter for unique values *)
pattern2Var = {Verbatim[Pattern][x_, Verbatim[_]] :> Var[x],
               Verbatim[_] :> Var[$, anon++] };
```

To understand these rules, recall that the internal form of $x\_$ is Pattern[x, Blank[]]. We want to treat x as a variable, so we wrap only the head Pattern in Verbatim. The pattern object Blank[] or $\_$, however, should be treated as a constant; it is, therefore, wrapped in Verbatim.

2.2 A PROLOG Interpreter for *Mathematica*

The result is `Verbatim[Pattern][x_, Verbatim[_]]`, the left side of the first rule. The `x_` inside it is now a pattern variable for *this* rule!

In close analogy to *Mathematica*'s downvalues, which are used to store definitions made for a symbol, logic rules shall be kept in a list `LogicValues[symbol]` attached (as an upvalue) to *symbol*.

The code for `Assert[lhs, rhs...]` appends the given logic rule given to the list already stored as `LogicValues` of the head of *lhs* (Listing 2.2–1).

```
Assert[lhs_, rhs___]  := insertRule[ makeRule[lhs, rhs], Append ]
Asserta[lhs_, rhs___] := insertRule[ makeRule[lhs, rhs], Prepend ]

insertRule[r_, op_] :=
    With[{h = dispatch[r]},
        If[ h === Null, Return[$Failed] ];
        LogicValues[h] ^= op[ LogicValues[h], r ];
    ]
makeRule[lhs_] := rule[lhs] /. Pattern2Var
makeRule[lhs_, True] := makeRule[lhs]
makeRule[lhs_, rhs_] := rule[lhs, rhs] /. Pattern2Var
makeRule[lhs_, rhs__] := makeRule[lhs, And[rhs]]
dispatch[rule[lhs_, ___]] := predicate[lhs]
predicate[(h_Symbol) | (h_Symbol[___])] := h
predicate[e_] := (Message[Logic::nohead, e]; Null)  (* no predicate *)
Format[rule[lhs_]] := lhs /. var2Pattern
Format[r:rule[_, _]] := Infix[ r /. var2Pattern, " :- " ]
LogicValues[_Symbol] := {}  (* for symbols without rules yet *)
```

Listing 2.2–1: Data type for rules

The auxiliary constructor `makeRule[lhs, rhs...]` takes care of rules without a right side (turning them into facts) and rules with more than one clause on the right side (wrapping And around the clauses). `rule` is our data type for facts and rules. A right side consisting of `True` can also be treated as a fact because it is always satisfied. The output form of rules approximates PROLOG's syntax. The function `predicate[expr]` returns the predicate to which the rule must be attached, either *expr* itself if it is a symbol, or the head of a composite *expr*. This definition is best expressed with an alternative. Downvalues in *Mathematica* are treated in the same way.

The last definition for `LogicValues` takes care of defining the first rule for a symbol. Without it the call to `LogicValues[h]` occurring in the code for `insertRule` would not return a proper value. This simple trick avoids a cumbersome test in the code. Because nonempty rule lists are stored as *upvalues*, the definition will apply only when there are indeed no rules yet.

`Assert` appends new rules at the end of the list of rules. `Asserta` prepends rules at the beginning. There is also a command `Retract[rule...]` that removes a rule from the database. The first rule matching the argument is removed.

2.2.1 Evaluating Predicates: State Information and Backtracking

A user-level query Query[*goal*] is transformed into internal form by converting the pattern variables as we have just seen. It is then handed to our internal query evaluator qeval.

A query is either a built-in predicate (treated in the next section) or a user-defined predicate for which we may have defined logic values.

The internal query evaluator is called as qeval[*goal, state*]. It takes the goal predicate and a *state* as arguments and returns two items: a list of bindings or $Failed if it cannot be satisfied, and the new state. The second argument of qeval describes the internal state of the evaluator, which is needed for backtracking. When the goal has to be retried, it is the caller's responsibility to call qeval[*goal, state*] with the state that was returned in the last call. The special value initialState is used in the first call and a value of finalState means that the goal cannot be retried.

Mathematica does no backtracking when applying rules. Therefore, its evaluator does not need any state information (the state is implicit in the run-time stack). Because backtracking is an essential feature of logic programming we need to manipulate the state in this complicated way.

If the goal is a user-defined predicate, we try all logic values in turn until we find one whose left side unifies with the goal (Listing 2.2–2). The state information needed to continue the search consists of the list of additional rules to try, the current rule, and its state. For efficiency, we keep a version of the current rule with its variables renamed. Renaming variables is necessary to avoid conflicts between variable names in rules and in queries.

```
(* apply rules: state is triplet {rest of rules, instance of first rule, its state} *)
qeval[ expr_, initialState ] :=
    With[{rules = LogicValues[predicate[expr]]},
        If[ Length[rules] == 0, Return[no] ];
        qeval[expr, {Rest[rules], rename[First[rules]], initialState}]
    ]

qeval[ expr_, {rules0_, inst0_, rstate0_} ] :=
    Module[{rules = rules0, inst = inst0, rstate = rstate0, res},
        While[True,
            res = tryRule[ expr, inst, rstate ];
            If[ !failedQ[res],
                Return[{bindings[res], {rules, inst, state[res]}}] ];
            (* else try next rule *)
            If[ Length[rules] == 0, Break[] ]; (* no more *)
            inst = rename[First[rules]];
            rstate = initialState;
            rules = Rest[rules];
        ];
        no
    ]
```

Listing 2.2–2: Searching for a rule

2.2 A PROLOG Interpreter for *Mathematica*

The loop is initialized by setting the list of remaining rules to the rest of the rules of the predicate, the current rule to the renamed version of the first rule, and its state to `initialState`. If there are no rules, we immediately return no which is defined as `{$Failed, finalState}`. The general case takes the current rule in the state information and tries it, passing the saved state information along. If this attempt succeeds, we return the bindings obtained and the updated state information. If it does not succeed, we have to try the next rule. If there are no more rules, we return with no. To try out a rule, we have to unify its left side with the goal, apply the bindings obtained to its right side, and call `qeval` recursively with the right side (see Listing 2.2–3).

```
(* for facts *)
tryRule[ expr_, r:rule[lhs_], initialState ] :=
    Module[{bind, inst, res},
        bind = Unify0[ expr, lhs ];        (* unify lhs *)
        If[ bind === $Failed, Return[no] ]; (* does not unify *)
        bind = closure[bind];
        {bind, finalState}
    ]

(* for rules proper *)
tryRule[ expr_, r:rule[lhs_, rhs_], initialState ] :=
    Module[{bind, res},
        bind = Unify0[ expr, lhs ];        (* unify lhs *)
        If[ bind === $Failed, Return[no] ]; (* does not unify *)
        tryRule[ expr, r, {bind, rhs //. bind, initialState} ]
    ]

tryRule[ expr_, r_rule, {bind_, inst_, stater_} ] :=
    Module[{res, outbind},
        res = qeval[ inst, stater ];        (* recursion *)
        If[ failedQ[res], Return[res] ];
        outbind = closure[bind, bindings[res]];
        {outbind, {bind, inst, state[res]}}
    ]
```

Listing 2.2–3: Applying a rule

The initial call does the unification. For efficiency, the result (the list of bindings) is retained as part of the state information, as is the instantiated right side. The first definition takes care of facts, that is, rules without a right side. The function `closure[`*bindings*`]` from **Unify.m** combines bindings. The actual code in **LogicProgramming.m** contains an array of additional details which we have no room here to discuss.

2.2.2 Built-In Predicates

The built-in predicates are similar to built-in functions and commands in *Mathematica*. They are used to do things that cannot be expressed by simple logic rules. Each built-in predicate p

is implemented by special code for `qeval[p[args...], state]`. This code is attached to p as an upvalue, which avoids a large number of (sequentially searched) rules for `qeval`. Note that we use the standard Boolean operators And, Or, and Not for our purpose.

Conjunction (And)

We have encountered one built-in predicate so far: $And[goal_1, goal_2, \ldots, goal_n]$ (most often written in infix form as $goal_1$ && $goal_2$ && $\ldots$ && $goal_n$). Evaluation happens by first (recursively) trying the first goal. If this goal succeeds, we apply the bindings obtained to the rest of the query $And[goal_2, \ldots, goal_n]$ and try it as well. If this query also succeeds, we return the combined bindings. If this second query fails, we retry the first one until it fails as well or until we find a success for the second query. The state information needed is the result of the first query and the state of the second query. For efficiency, we keep also the instance of the rest of the query (Listing 2.2–4).

Because of the state information, the recursive treatment of the rest of the query is simpler to implement than is a loop over all arguments of And. The same kind of *tail recursion* instead of iteration is used in LISP.

```
(* And: state is triplet {result of first clause, instance of rest, rest state} *)
And/: qeval[ and:And[g1_, goals__], initialState ] :=
   Module[{res},
       (* call first goal to initialize things *)
       res = qeval[g1, initialState];
       If[ failedQ[res], Return[res] ]; (* failure *)
       (* apply bindings to rest and call it *)
       qeval[ and, {res, And[goals] //. bindings[res], initialState} ]
   ]
And/: qeval[ And[g1_, goals__], {res10_, rest0_, stater0_} ] :=
   Module[{res1 = res10, rest = rest0, stater = stater0, res, binds},
       While[ True,
           res = qeval[ rest, stater ]; (* eval rest *)
           If[ !failedQ[res],
               binds = closure[bindings[res1], bindings[res]];
               Return[{binds, {res1, rest, state[res]}}];
           ];
           (* else try first one again *)
           res1 = qeval[g1, state[res1]];
           If[ failedQ[res1], Return[res1] ];
           stater = initialState; (* reset for next attempt *)
           rest = And[goals] //. bindings[res1];
       ];
   ]
```

Listing 2.2–4: Evaluating a conjunction of goals

Disjunction (Or)

The disjunction $goal_1 \,||\, goal_2 \,||\, \ldots \,||\, goal_n$ is evaluated as follows. The first goal is called. If it succeeds, we are done. If it fails, we discard it and try the rest of the disjunction $goal_2 \,||\, \ldots \,||\, goal_n$ until we find a goal that succeeds. To redo a disjunction, we try to redo the goal that succeeded last time. The state information consists simply of the state of the first goal and the state of the rest (Listing 2.2–5).

```
(* Or: state is  pair {firststate, nextstate} *)
Or/: qeval[ or_Or, initialState ] :=
    qeval[ or, {initialState, initialState} ]
Or/: qeval[ Or[g1_, goals__], {state1_, stater_} ] :=
    Module[{res},
        (* try first clause *)
        res = qeval[g1, state1];
        If[ !failedQ[res],
            Return[{bindings[res], {state[res], initialState}}] ];
        If[ cutQ[res], Return[res] ]; (* no alternatives in this case *)
        (* try rest of them *)
        res = qeval[ Or[goals], stater ];
        If[ failedQ[res], Return[res] ];
        {bindings[res], {finalState, state[res]}} (* don't try first one again *)
    ]
```

Listing 2.2–5: Disjunction of goals

This rule says that an *edge* is an undirected arc, which means that there is an arc either from x to y or one in the opposite direction.

```
In[10]:= Assert[ edge[x_, y_], arc[x_, y_] || arc[y_, x_] ]
```

Here are all edges having c as one of their endpoints.

```
In[11]:= QueryAll[ edge[c, v_] ]
{v -> d}
{v -> a}
```

Negation (Not)

The goal Not[*goal*] or !*goal*, succeeds if *goal* itself fails. No bindings are generated and it cannot be redone. Its implementation is therefore quite simple (see Listing 2.2–6).

```
Not/: qeval[Not[g_], initialState ] :=
    If[ failedQ[ qeval[g, initialState] ], yes, no ]
```

Listing 2.2–6: Negation as failure

Negation provides the best illustration that logic programming is not mathematical logic. In logic programming we interpret negation as failure. Here is the standard example:

Logically, this result would mean that "Mary is not human." In logic programming, we interpret negation as failure, so it really means "there is not enough information in the database to prove 'Mary is human'."

```
In[12]:= Query[ !human[Mary] ]
Out[12]= Yes
```

This use of Not is called the *closed world assumption*. We assume that our database is complete and anything not deducible from it is, therefore, false.

Equality and Inequality

In logic programming, we interpret equality as unification and inequality as nonunification. We have already seen an example of inequality in the query path[a, v_] && v_ != a in the introduction. Because *Mathematica* turns Not[a == b] into a != b, we need the definition for inequality, even though it is equivalent to Not[a == b]. Unification cannot be redone. The code is shown in Listing 2.2–7.

```
Equal/: qeval[ e1_ == e2_, initialState ] :=
   Module[{u},
       u = Unify0[e1, e2];
       If[ u === $Failed, no, {u, finalState} ]
   ]
Unequal/: qeval[ e1_ != e2_, initialState ] :=
       If[ Unify0[e1, e2] === $Failed, yes, no ]
```

Listing 2.2–7: Equality is unification

Input and Output

It is hard to fit input and output into the logic framework. PROLOG suffers from the *single-paradigm syndrome:* everything has to be forced into the single programming paradigm advocated by the language, even if it does not fit naturally. We provide only rudimentary I/O facilities (different from the ones found in PROLOG). input[v, *prompt*] reads an input expression using the (optional) prompt and succeeds if the input read can be unified with v. Most often, v is a variable and unification always succeeds. This case corresponds roughly to *Mathematica*'s v=Input[*prompt*]. If the end-of-file character (CTRL-D) is encountered, it fails. input can be redone; it simply asks for new input.

2.2 A PROLOG Interpreter for *Mathematica*

The "predicate" print[*args*...] prints its arguments (using Print[*args*...]) and always succeeds. It cannot be redone (Listing 2.2–8).

```
input/: qeval[ input[ v_, prompt_:"? " ], initialState ] :=
    Module[{in},
        in = Input[prompt] /. pattern2Var;
        If[ in === EndOfFile,
            no,
            { Unify0[v, in], initialState } (* can be redone *)
        ]
    ]
print/: qeval[ print[args___], initialState ] := ( Print[args]; yes )
```

Listing 2.2–8: Input and output

This loop repeatedly asks for a vertex and prints yes if there is a path from a to it, and no otherwise. The final fail is a predicate that always fails. It forces backtracking and thus implements the loop.

```
In[13]:= Query[ input[v_, "vertex: "],
                path[a, v_] && print["yes"] ||
                    !path[a, v_] && print["no"],
                fail ]
vertex: a
yes
vertex: e
no
vertex: b
yes
vertex: ^D

Out[13]= No
```

Procedural Interpretation of Logic Rules

In pure logic, the predicates True and False would not alter the meaning of a program. The expression *p* && True can be simplified to *p* (*Mathematica* does this). Because a logic program has—besides its declarative interpretation—a procedural meaning, defined by the actions of the query evaluator, such predicates can alter the behavior of the program. The predicate true always succeeds. We provide it (in addition to True) because it is not affected by the built-in simplification above. The predicates fail and false always fail. Their implementation is trivial: any undefined predicate will fail. For efficiency reasons, we nevertheless give a definition for them (see Listing 2.2–9).

This is the skeleton of a PROLOG-level implementation of QueryAll. The final false forces backtracking until all possibilities are exhausted.

```
In[14]:= Query[path[a, v_], print[v_], false]
a
b
c
d

Out[14]= No
```

```
qeval[ True, initialState ] := yes
qeval[ False,initialState ] := no
fail/: qeval[ fail, initialState ] := no
true/: qeval[ true, initialState ] := yes
false = fail

Assert[ repeat ]
Assert[ repeat, repeat ]
```

Listing 2.2–9: Procedural version of logic constants

It does not work with the built-in False, because *Mathematica* simplifies And[*goals...*, False] to False.

```
In[15]:= Query[path[a, v_], print[v_], False]
Out[15]= No
```

The predicate `repeat` succeeds (like `true`), but it can be redone infinitely often. It can be used to form loops. It can be implemented directly as a logic rule (Listing 2.2–9).

The Cut

There is one remaining predicate that is the topic of endless discussions: the *cut*. The cut prevents backtracking and is, therefore, not part of a pure logic programming language, but it seems unavoidable and often leads to big gains in program efficiency. (This might be viewed as a corollary to the single-paradigm syndrome: every such language has a feature that breaks its single paradigm.) If a cut is encountered on the right side of a rule, no backtracking will occur to the left of the cut. If the cut should be redone, the current goal fails immediately. There is no attempt to redo the predicates before the cut, nor does it try a different rule for the same predicate! A cut, therefore, commits the choices made so far.

One use is for programming mutually exclusive alternatives. The conditional definition

$$f[x\_] := If[pred[x], then[x], else[x]]$$

needs to be programmed like this (using PROLOG syntax):

$$f[X] :- pred[X], then[X].$$
$$f[X] :- !pred[X], else[X].$$

The negation of the predicate in the else clause is necessary to avoid backtracking into the second rule if the predicate is true. With a cut, we can commit ourselves to the first clause, once the predicate is known to be true:

$$f[X] :- pred[X], cut, then[X].$$
$$f[X] :- else[X].$$

Negation can also be implemented in terms of the cut as

```
not[P] :- (P && cut && fail) || true.
```

The cut prevents backtracking from `fail`, once P is known to be true.

The cut is implemented through the use of a special state that acts like a failure state but is detected by the code for And, Or, and the loop that tries rules in turn. When the cut state is encountered, no further attempts to redo the goal are made and failure is returned.

Equation Solving

In PROLOG, the special predicate `is[var = expr]` is used to evaluate arithmetic expressions. Because all queries in our logic evaluator are evaluated by *Mathematica* in the standard way, we do not need it for this purpose. We use `is[eqlist]` to tap some of *Mathematica*'s power. This predicate succeeds if the occurring variables can be bound to a solution of the list of equations *eqlist* (using Solve). If this query is redone, the next solution is returned, and so on. Implementation is quite simple, because Solve already returns lists of replacements for the variables. The state information is simply the list of remaining solutions (see Listing 2.2–10).

```
(* equation solving: state is list of remaining solutions *)
is/: qeval[ t:is[ eq_ ] , initialState ] := qeval[ t, Solve[ eq, variables[eq] ] ]
is/: qeval[ _is , {} ] := no    (* no more solutions *)
is/: qeval[ _is , sols_List ] := {First[sols], Rest[sols]}
```

Listing 2.2–10: Solving equations

This query returns the first solution of the quadratic equation.

$$\text{In[16]}:= \text{Query}[\text{ is}[\text{ x\_}^2 == \text{x\_} + 1 \text{]]}$$

$$\text{Out[16]}= \{x \to \frac{1 - \text{Sqrt}[5]}{2}\}$$

Now we get the second solution.

$$\text{In[17]}:= \text{Again[]}$$

$$\text{Out[17]}= \{x \to \frac{1 + \text{Sqrt}[5]}{2}\}$$

Meta-logic Predicates

As in most other languages, there are some predicates that leave the primary domain of the language (logic, in our case). Examples are `explode` in LISP, which turns an atom into a list of characters (like `Characters[ToString[`*symbol*`]]` in *Mathematica*), and `DownValues[`*symbol*`]` in *Mathematica*, which allows us to mess around with definitions. In PROLOG, there is `assert[`*rule*`]`, which works like the outside `Assert[`*rule*`]`; `retract[`*rule*`]`, corresponding to `Retract[`*rule*`]`;

var[v], which is true if v is an uninstantiated variable; ground[$expr$], which tells whether *expr* is free of variables (a so-called *ground term*); and name[v, *list*], which succeeds if *list* is the list of character codes that make up the symbol v.

name can be used to find the character codes of the name of a symbol.	In[18]:= Query[name[symbol, l_]] Out[18]= {l -> (115 121 109 98 111 108)}
In typical PROLOG manner, it can also be used "backward" to assemble a list of character codes into a symbol.	In[19]:= Query[name[var_, list[97, 108, 112, 104, 97]]] Out[19]= {var -> alpha}

The ability to modify the database of logic rules during a query with assert[*rule*] can be used for dynamic programming, similar to what we can do in *Mathematica*. Here is the example for Fibonacci numbers, taken from Chapter 7 of the first volume of this book. This is the usual recursive program for Fibonacci numbers in PROLOG:

```
Assert[ fib[1, 1] ]
Assert[ fib[2, 1] ]
Assert[ fib[n_, f_], n_ > 2, n1_ == n_ - 1, fib[n1_, f1_],
              n2_ == n_ - 2, fib[n2_, f2_], f_ == f1_ + f2_ ]
```

It shows the expected poor performance. If we store any newly computed values as new logic rules, they will not have to be recomputed. The new rules need to go to the beginning of the list, so we use asserta:

```
Assert[ fibf[1, 1] ]
Assert[ fibf[2, 1] ]
Assert[ fibf[n_, f_], n_ > 2, n1_ == n_ - 1, fibf[n1_, f1_],
              n2_ == n_ - 2, fibf[n2_, f2_], f_ == f1_ + f2_,
              asserta[fibf[n_, f_]] ]
```

This query computes the tenth Fibonacci number.	In[1]:= Query[fibf[10, f_]] Out[1]= {f -> 55}

We can see that all intermediate values have been stored.	In[2]:= LogicValues[fibf]
	Out[2]= {fibf[10, 55], fibf[9, 34], fibf[8, 21],
	  fibf[7, 13], fibf[6, 8], fibf[5, 5], fibf[4, 3],
	  fibf[3, 2], fibf[1, 1], fibf[2, 1],
	  fibf[n_, f_] :- ((n_) > 2 && (n1_) == -1 + (n_) &&
	    fibf[n1_, f1_] && (n2_) == -2 + (n_) &&
	    fibf[n2_, f2_] && (f_) == (f1_) + (f2_) &&
	    asserta[fibf[n_, f_]])}

2.2 A PROLOG Interpreter for *Mathematica*

In *Mathematica* we would use a definition of the form

$$\text{fibf[n\_] := fibf[n] = fibf[n-1] + fibf[n-2]}.$$

When retracting rules, you can use unification to search for the rule to be removed. If there are several matching rules, backtracking will remove one after another.

These are the rules known for `arc`.	`In[1]:= LogicValues[ arc ]` `Out[1]= {arc[a, b], arc[a, c], arc[c, d]}`
This query succeeds because a matching rule was found and removed.	`In[2]:= Query[ retract[arc[c, d]] ]` `Out[2]= Yes`
It is indeed gone.	`In[3]:= LogicValues[ arc ]` `Out[3]= {arc[a, b], arc[a, c]}`

This query removes all rules of the form `arc[a, `x`]`. There were two of them.	`In[4]:= QueryAll[ retract[arc[a, x_]] ]`
	`{x -> b}`
	`{x -> c}`

Here, we remove all rules for the predicate `path` that are not facts (that do have a right side). Note that you do not need to use the same variable names as in the rule; unification takes care of that.	`In[5]:= QueryAll[ retract[path[u_, v_], rhs_] ]` `{rhs -> arc[u, y] && path[y, v]}`

Table 2.2–1 shows all predicates defined in our interpreter. These are the predicates present in most PROLOG interpreters. This table is our reference manual.

2.2.3 Running the Evaluator

If you want to experiment with logic queries, you can read in the package LogicProgramming.m and then set up some logic rules with `Assert`. You have already seen how to formulate queries. `Again` is implemented by storing the state returned from the internal evaluator in a static private variable. While developing your programs, you can clear all existing logic values with `Clear[`*symbol*`]` in the same way that you would clear *Mathematica* definitions before entering them again.

Table 2.2–2 shows the commands available for working with the logic query evaluator. Please note that all assertions and queries are still evaluated by *Mathematica* in the usual way.

The variable `prolog` is bound to a procedure that simulates a PROLOG-style main loop. In this loop, you can enter queries directly at the prompt `?-`.

You can read in one of the examples directly. The interpreter is loaded automatically.	`In[1]:= << MathProg/DAG.m`

`true`	always succeeds
`false, fail`	always fails
`cut`	succeeds but prevents backtracking
`repeat`	succeeds and can be redone always
`And[`*goals*`]`, g_1`&&`g_2`&&`...	succeeds if the goals succeed in sequence
`Or[`*goals*`]`, g_1`||`g_2`||`...	succeeds if one of the goals succeeds
`Not[`*goal*`]`, `!`*goal*	succeeds if *goal* fails
e_1 `==` e_2	e_1 and e_2 can be unified
e_1 `!=` e_2	e_1 and e_2 cannot be unified
`print[`*args*`]`	prints the *args* and succeeds
`input[`v`, (`*prompt*`)]`	unifies v with an expression read
`name[`v`, `l`]`	l is the list of character codes of the symbol v
`var[`v`]`	v is an uninstantiated variable
`nonvar[`v`]`	v is not a variable
`ground[`e`]`	e contains no variables
`integer[`i`]`	i is an integer
`atomic[`a`]`	a is an atom
`atom[`a`]`	a is an atom other than a number
`is[`*eqlist*`]`	succeeds if the equations *eqlist* can be solved
`assert[`*rule*`]`	succeeds if the rule was entered into database
`asserta[`*rule*`]`	succeeds if the rule was entered at the beginning
`assertz[`*rule*`]`	succeeds if the rule was entered at the end
`retract[`*rule*`]`	succeeds if the rule was removed from the database
`run[`*cmd*`]`	evaluates the *Mathematica* expression *cmd* and succeeds if no messages were produced
`run[`*cmd*`, `*res*`]`	succeeds if the result of *cmd* is *res*
`trace`	always succeeds and turns on full tracing
`notrace`	always succeeds and turns off full tracing

Table 2.2–1: The built-in predicates of our logic language

2.2 A PROLOG Interpreter for *Mathematica*

Assert[*fact*]	assert *fact*.
Assert[*lhs*, g_1, ..., g_n]	assert *lhs* :- g_1, ..., g_n.
Asserta[*rule*]	enter *rule* at the beginning
Assertz[*rule*]	enter *rule* at the end
Retract[*rule*]	remove *rule*
LogicValues[*symbol*]	the list of logic rules for *symbol*
Clear[*symbol*]	remove all logic rules defined for *symbol*
Query[g_1, ..., g_n]	evaluate the query ?- g_1, ..., g_n.
QueryAll[g_1, ..., g_n]	print all solutions
Again[]	redo the last query
Spy[*symbol*]	turn on tracing for rules attached to *symbol*
NoSpy[*symbol*]	turn off tracing for *symbol*
NoSpy[]	turn off tracing for all symbols
traceLevel	trace level (initially 1); 0 disables all tracing, 2 shows details
Yes	indicates a successful query without variables
No	indicates an unsuccessful query
{var_1->val_1, ..., var_n->val_n}	bindings that satisfy a query

Table 2.2–2: Commands and values of the query evaluator

After each result you are prompted for redoing the query. If you enter ; the query is redone; if you enter RETURN it is abandoned. To get out of the query loop, enter an end-of-file character (^D).

```
In[2]:= prolog

?- arc[a, x_]
{x -> b} ?;
{x -> c} ?;
No

?- path[a, x_]
{x -> a} ?;
{x -> b} ?

?- ^D
```

2.2.4 Debugging

Because the flow of control is not obvious in a logic program, good debugging facilities are essential. Tracing is the most useful of these. It shows which goals are attempted and with what bindings they succeed. Spy[*symbol*] turns on tracing for rules attached to *symbol* (like *Mathematica*'s On[*symbol*]). NoSpy[*symbol*] turns tracing off. NoSpy[] turns it off for all symbols.

Let us look again at the examples with the DAG.

```
In[3]:= Spy[path]
```

Now we see all calls to the procedure path.

```
In[4]:= Query[ path[a, x_] ]
    Goal is path[a, x_]
    Yes, with {x -> a}

Out[4]= {x -> a}
```

When we redo the query, the next rule is tried.

```
In[5]:= Again[]
    Goal is path[a, x_]
      Goal is path[b, x_]
      Yes, with {x -> b}
    Yes, with {x -> b}

Out[5]= {x -> b}
```

If you set traceLevel to 2, you will also see all rules that are tried, together with the instantiated subqueries.

```
In[6]:= traceLevel=2;
```

In this form of output, variables are given in their internal form. The suffixes distinguish the variables from different invocations of the same rule.

```
In[7]:= Again[]
    Goal is path[a, x_]
      >redo: path[x_, z_] :- (arc[x_, y_] && path[y_, z_])
      +new goal: arc[a, y ] && path[y , x_]
                         1              1
        Goal is path[b, x_]
          >call: path[x_, z_] :- (arc[x_, y_] && path[y_, z_])
          +new goal: arc[b, y ] && path[y , x_]
                             4              4
          <failed
        No.
        Goal is path[c, x_]
          >call: path[x_, x_]
          <ret:  {x  -> c, x_ -> c}
                     9
        Yes, with {x -> c}
      <ret:  {x  -> a, z  -> c, y  -> c, x  -> c, x_ -> c}
                 1         1        1        9
    Yes, with {x -> c}

Out[7]= {x -> c}
```

2.3 Lists in PROLOG

PROLOG uses the same representation of lists as does LISP. A list consists of a first element, the *head*, and the rest of the elements, the *tail*. A typical pattern in PROLOG is [H|T], standing for the list with first element H and rest T. In *Mathematica*, we would write this pattern as {h_, t___} and then use {t} for the rest (t itself is only the *sequence* of elements, so we have to wrap it in a list). Because our simple unifier cannot handle sequences, we use LISP lists, imported from the package Lisp.m (developed in the first volume, Chapter 2).

2.3 Lists in PROLOG

In this notation, we use cons[h_, t_] for the list with first element h and rest t; car[l] is the first element of l, and cdr[l] is the rest of l. nil is the empty list. list[e_1, e_2, ..., e_n] generates the list (e_1 e_2 ... e_n).

In this section we shall have a look at some typical list-processing programs. The code of all the examples is collected in Lists.m.

2.3.1 Sorting Lists

Here is a predicate that expresses that a list is ordered.

```
Assert[ ordered[list[]] ]
Assert[ ordered[list[_]] ]
Assert[ ordered[cons[e_, r_]], e_ < car[r_], ordered[r_] ]
```

The empty list is always ordered, as is any list with one element. If the list has at least two elements, it is ordered if its first element is smaller than the first element of its rest and the rest is also ordered.

2.3.2 Joining Lists

In *Mathematica* and in LISP, there is a function Join[*list₁*, *list₂*] that joins two lists. In PROLOG, such functions are expressed as predicates, where the desired result is an additional argument. Therefore, the predicate join[] will have three arguments. The predicate join[*list₁*, *list₂*, *list*] expresses: "the join of *list₁* and *list₂* is *list*." Let us find rules that characterize this predicate.

The join of the empty list and *l* is *l*.	In[1]:= Assert[join[nil, l_, l_]]
The join of the list with first element *e* and rest *r* and a list *l* is the list with the same first element *e* and rest *s*, provided that the join of *r* and *l* is equal to *s*.	In[2]:= Assert[join[cons[e_, r_], l_, cons[e_, s_]], join[r_, l_, s_]]

That's all! We have not really written a program to compute the join of two lists. But our predicate works and it can do more than compute the join of two lists.

This query computes the join of (a b) and (c d) and "returns" the result as a binding for the one variable present. This form resembles an ordinary function call res = join[(a b), (c d)].	In[3]:= Query[join[list[a, b], list[c, d], res_]] Out[3]= {res -> (a b c d)}

This query asks: "which list, when joined to the list (a b), gives (a b c d)?"

```
In[4]:= Query[ join[list[a, b], l2_, list[a, b, c, d]] ]
Out[4]= {l2 -> (c d)}
```

This query finds all possible ways to join two lists to give (a b c d).

```
In[5]:= QueryAll[ join[l1_, l2_, list[a, b, c, d]] ]
{l1 -> (), l2 -> (a b c d)}
{l1 -> (a), l2 -> (b c d)}
{l1 -> (a b), l2 -> (c d)}
{l1 -> (a b c), l2 -> (d)}
{l1 -> (a b c d), l2 -> ()}
```

As you can see, there is no distinction between input and output parameters. Not all PROLOG programs are that flexible, but it is a good idea to try to write code that can be used this way, if possible. Often, arithmetic inequalities prevent the generation of all possibilities. In our first example (the predicate ordered[]), we unfortunately cannot use Query[ordered[l_]] to enumerate all sorted lists. There is an infinite number of them, anyway.

2.3.3 Testing Membership

Here is one last list example. We define a predicate member[e, l] that says "e is a member of list l."

The element e is a member of the list whose first element is e.

```
In[6]:= Assert[ member[e_, cons[e_, _]] ]
```

The element e is a member of the list with rest r, provided it is an element of r.

```
In[7]:= Assert[ member[e_, cons[_, r_]], member[e_, r_] ]
```

We find that a is indeed a member of (a b c a).

```
In[8]:= Query[ member[a, list[a, b, c, a]] ]
Out[8]= Yes
```

It should now come as no surprise that we can invert the question to find an element that is a member of (a b c a).

```
In[9]:= Query[ member[e_, list[a, b, c, a]] ]
Out[9]= {e -> a}
```

Our program has a bug: elements that occur more than once are listed more than once.

```
In[10]:= QueryAll[ member[e_, list[a, b, c, a]] ]
{e -> a}
{e -> b}
{e -> c}
{e -> a}
```

Here is a way to correct this problem. The second rule succeeds only if the candidate element is different from the first element of the list. Note that the inequality must come last.

```
In[11]:= Assert[ member2[x_, cons[x_, l_]] ];\
         Assert[ member2[x_, cons[y_, l_]],
                 member2[x_, l_], x_ != y_ ]
```

2.4 Backtracking

Each element that appears is now listed only once.

```
In[12]:= QueryAll[ member2[e_, list[a, b, c, a]] ]
{e -> a}
{e -> b}
{e -> c}
```

If we give up the requirement that each argument can be used for input or output, we can find a simpler definition to test membership by using a cut as soon as an element has been found.

```
In[13]:= Assert[ member1[x_, cons[x_, l_]], cut ];\
         Assert[ member1[x_, cons[_, l_]], member1[x_, l_] ]
```

2.4 Backtracking

Backtracking is a method for exhaustive search or depth-first visit of a tree. Whenever you reach a point where there is no way to proceed (such as the leaf of a tree), you retrace your steps to the last point where you had a choice of paths. Backtracking is the method used by the query evaluator in an attempt to satisfy or resatisfy a goal. Therefore, problems involving exhaustive search can be implemented in PROLOG easily. Let us look at two examples: nondeterministic automata and games.

2.4.1 Nondeterministic Automata

In a logic program, we can give several alternative rules for a predicate and try them out in some unspecified order; if one does not succeed, we backtrack and try out another one. This possibility makes it easy to model nondeterministic processes, for example, a *nondeterministic finite automaton*.

A finite automaton (FSA) consists of a number of *states,* usually represented as the vertices of a graph. *State transitions* lead from one state to another one, represented as arcs between states. A state transition is triggered by an input symbol. Each arc is therefore labeled by a symbol and only the transition labeled with the current input symbol can be taken. The current input symbol is "used up" by the transition and then the next one is looked at. When the input is exhausted, the automaton halts. A subset of states is designated as *final.* If the automaton halts in a final state, it *accepts* the input; otherwise, it *rejects* it. If no arc labeled with the current input symbol exists in the current state, the computation also terminates and rejects. This automaton is just like a Turing machine without the tape. (See Chapter 5.)

An automaton is *deterministic* if there is at most one arc with each label starting at each state. In this case, there is never a choice of which transition to take; at most one transition is possible. If this property is not satisfied, the automaton is nondeterministic. Such automata may also allow a different kind of transition: *silent transitions,* which are not labeled and can be taken without using up an input symbol. If there is at least one computation that halts in a final state, a nondeterministic automaton accepts. Because there can be several choices for the continuation of a computation, there can be several states in which the automaton halts for the same input. Because we must answer the question regarding whether *one* accepting computation exists, we have to perform an exhaustive search through all possible computations. PROLOG can do this with a minimum of programming; we need only three rules!

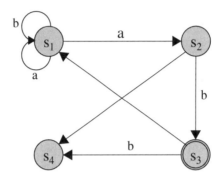

Figure 2.4–1: A small nondeterministic finite automaton

An automaton is represented as a number of facts that describe the possible transitions and silent transitions. The predicate trans[s_1, i, s_2] describes a transition from state s_1 to state s_2 labeled with symbol i, silent[s_1, s_2] describes a silent transition from state s_1 to state s_2, and final[s] designates state s as final. The automaton from Figure 2.4–1 is coded as shown in Listing 2.4–1.

```
Assert[ trans[s1, a, s1] ]
Assert[ trans[s1, a, s2] ]
Assert[ trans[s1, b, s1] ]
Assert[ trans[s2, b, s3] ]
Assert[ trans[s3, b, s4] ]

Assert[ silent[s2, s4] ]
Assert[ silent[s3, s1] ]

Assert[ final[s3] ]
```

Listing 2.4–1: FSA.m: The automaton from Figure 2.4–1 as a logic program

Now, we want to define the predicate accepts[s, *input*] that says the automaton accepts the input (given as a list of symbols), starting in state s. The rules are shown in Listing 2.4–2.

The first rule says that the empty input is accepted if we are in a final state. The second rule says that the input starting with symbol x is accepted if we perform a transition labeled with x leading to a new state s_1 and then accept the rest of the input in this state. The third rule handles silent transitions that do not modify the input in any way.

2.4 Backtracking

Assert[accepts[s_, nil], final[s_]]

Assert[accepts[s_, cons[x_, rest_]], trans[s_, x_, s1_], accepts[s1_, rest_]]

Assert[accepts[s_, input_], silent[s_, s1_], accepts[s1_, input_]]

Listing 2.4–2: The rules for nondeterministic automata

This query shows that the input aaab is accepted from state s1.	`In[1]:= Query[ accepts[s1, list[a, a, a, b]] ]` `Out[1]= Yes`

Here, we find all states from which the input ab is accepted.	`In[2]:= QueryAll[ accepts[s, list[a, b]] ]`
	`{s -> s1}`
	`{s -> s3}`

Here are all inputs of length three that are accepted from state s1.	`In[3]:= QueryAll[ accepts[s1, list[x1_, x2_, x3_]] ]`
	`{x1 -> a, x2 -> a, x3 -> b}`
	`{x1 -> b, x2 -> a, x3 -> b}`

If we want to see the accepting inputs as lists, we can formulate the query this way.	`In[4]:= QueryAll[`
	`           input_ == list[_, _, _], accepts[s1, input_]`
	`       ]`
	`{input -> (a a b)}`
	`{input -> (b a b)}`

This example shows how flexible logic programming is and how easy it is to solve tasks that involve searching a large number of cases.

2.4.2 Games

Games are a good example of backtracking and searching. In a finite two-person game with complete information, the outcome can in principle be computed by traversing the whole game tree. For interesting games, this tree is much too large for an exhaustive search. Other methods are used to approximate exhaustive search, such as static evaluation of nonfinal positions to cut the search depth and *alpha-beta pruning* to cut the search breadth. PROLOG programs for these techniques are discussed in [8].

Here, we show how exhaustive search is programmed for the game of NIM. This simple game can be solved by other means, but we shall ignore them here. (For an excellent description of the mathematical theory behind NIM and many other games, see [6].)

The game of NIM involves a number of *heaps*, each consisting of a number of *tokens*. Each player moves in turn. A move consists of taking any number (at least one) of tokens from one heap. If all tokens in a heap are taken, the heap vanishes. If there are no more heaps left, you cannot move and you lose. This means that you win if you take the last token.

We shall describe a position in the game by a list of the numbers of tokens in the heaps. Thus, (3 4 5) describes a position consisting of three heaps with 3, 4, and 5 tokens, respectively. A move is described by the predicate `take[heap, amount]`. It describes the number of the heap from which the tokens are taken and the number of tokens removed. The predicate play[*pos₁*, *move*, *pos₂*] expresses the relation that *pos₂* is obtained from *pos₁* by performing the move *move*. Finally, win[*pos*, *move*] says that we can win in position *pos* by performing the move *move*.

The whole strategy of the game is expressed in a single rule:

Assert[win[pos_, t_], play[pos_, t_, pos1_], !win[pos1_, _]]

which says that you can win in position *pos* with move *t* if it is impossible to win in the resulting position *pos₁*. Because it is then the opponent's turn, this implies that you win!

The rules for play are a bit tricky. It would be a simple matter to find rules that could be used to compute the final position if both the initial position and the move were given, as in play[list[3, 4, 5], take[1, 2], pos1_], but we have to make sure the predicate works also with other arguments uninstantiated. In particular, we want to use play[list[3, 4, 5], t_, pos1_] to *generate* all possible moves in a given position. Listing 2.4–3 shows a set of rules that works.

```
(* take all *)
Assert[ play[ cons[n_, rest_], take[1, n_], rest_ ] ]
(* take one *)
Assert[ play[cons[n_, rest_], take[1, 1], cons[n1_, rest_]], n_ > 1, n1_ == n_ - 1 ]
(* one more than a move with smaller heap *)
Assert[ play[cons[n_, rest_], take[1, m_], cons[nr_, rest_]],
        n_ > 2, n1_ == n_-1,
        play[cons[n1_, rest_], take[1, m1_], cons[nr_, rest_]], m_ == m1_+1 ]
(* move in rest of heaps *)
Assert[ play[cons[n_, rest_], take[i_, m_], cons[n_, rest1_]],
        play[rest_, take[j_, m_], rest1_], i_ == j_+1 ]
```

Listing 2.4–3: Part of NIM.m: Move generation for NIM

The first three rules all generate or perform moves affecting the first heap. In the first rule, we take all tokens in the first heap, which makes it disappear. In the second rule, we take one token of the first heap, which consists of at least two tokens (so it does not go away). In the third rule, we use recursion. We generate a move in a position in which the first heap has one token *less* than in the given position. If we then take one token *more* than in this move, we arrive at the same final position. This rule applies if there are more than two tokens in the first heap. The last rule performs the moves in the rest of the heaps. All we need to do is add one to the number of the heap from which the tokens are taken.

This query performs a move by figuring out the position after the move.

In[1]:= Query[play[list[3, 4, 5], take[1, 2], p1_]]

Out[1]= {p1 -> (1 4 5)}

2.5 Deduction

Here is a list of all possible moves from position (1 2) and the resulting positions after the move.

```
In[2]:= QueryAll[ play[list[1, 2], t_, p_] ]
{t -> take[1, 1], p -> (2)}
{t -> take[2, 2], p -> (1)}
{t -> take[2, 1], p -> (1 1)}
```

We can ask for the move that transforms (1 2) into (2).

```
In[3]:= Query[ play[list[1, 2], t_, list[2]] ]
Out[3]= {t -> take[1, 1]}
```

Here is the list of all moves from (1 2 3) that do not remove any of the heaps completely (we request the new position to still have three heaps).

```
In[4]:= QueryAll[ play[list[1, 2, 3], t_, list[_, _, _]] ]
{t -> take[2, 1]}
{t -> take[3, 1]}
{t -> take[3, 2]}
```

No move can be made from the empty position. There is no rule expressing this fact, but all rules for play require the initial position to be nonempty.

```
In[5]:= Query[ play[nil, t_, p_] ]
Out[5]= No
```

This query shows that we can win from position (2) by taking all tokens.

```
In[6]:= Query[ win[list[2], t_] ]
Out[6]= {t -> take[1, 2]}
```

The position (1 1) is a *winning position*: If we achieve it after our move, we can win because there is no winning move for the opponent from this position.

```
In[7]:= Query[ win[list[1, 1], t_] ]
Out[7]= No
```

This query computes the winning move and the resulting position from position (1 2 1).

```
In[8]:= Query[ pos_==list[1,2,1],
               win[pos_, t_],
               play[pos_, t_, newpos_] ]
Out[8]= {pos -> (1 2 1), t -> take[2, 2], newpos -> (1 1)}
```

There is no winning move from the new position.

```
In[9]:= Query[ win[newpos /. %, t_] ]
Out[9]= No
```

Exhaustive search is a slow process. Even computing the winning move from position (3 4 5) takes approximately an hour of computing time.

2.5 Deduction

One of the main intended applications of logic programming was automatic theorem proving. We shall explain the method PROLOG uses and then given an application: deductive databases and expert systems.

2.5.1 Theorem Proving

The theoretical foundation of the method used by PROLOG for query evaluation is *resolution*, a technique developed for automatic theorem proving. The language of theorem proving is *predicate calculus*. We shall denote predicates by variables p, q, and r. The logic operators can all be expressed in terms of and ($\wedge$), or ($\vee$), and not ($\neg$). For example, the implication

$$p_1 \wedge p_2 \wedge \ldots \wedge p_n \to q \qquad (2.5\text{--}1)$$

is equivalent to

$$\neg(p_1 \wedge p_2 \wedge \ldots \wedge p_n) \vee q \qquad (2.5\text{--}2)$$

and, finally, to

$$\neg p_1 \vee \neg p_2 \vee \ldots \vee \neg p_n \vee q. \qquad (2.5\text{--}3)$$

A *literal* is either a predicate symbol or a negated predicate symbol. A disjunction (or connection) of literals is called a *clause*. The *satisfiability problem* consists of a set of clauses, understood to be the conjunction (and connection) of the clauses. The question to be answered is whether the predicate symbols have an assignment of truth values that makes the set of clauses true. For this to be the case, there must be at least one literal in each clause that has the value True assigned. All clauses are then true (they are disjunctions).

The method of resolution consists of transforming a set of clauses into another, simpler, but equivalent, set of clauses. If we ever generate an empty clause, we know that the set cannot be satisfied, because there is no literal in this clause that could be assigned true. If we restrict the clauses to *Horn clauses*, such transformations can be derived easily. A Horn clause is a clause that contains at most one nonnegated (positive) literal. Note that the clause derived from the implication 2.5–1 is a Horn clause. Axioms, which consist of just one positive literal, are also Horn clauses.

The resolution algorithm looks for two clauses of the forms $\{p, q\}$ and $\{\neg p, r\}$. These two can be replaced by the single clause $\{q, r\}$, as can be seen by looking at the two possible cases. The literal p must either be assigned true or false. If it assigned true, the second clause can only be satisfied if r is assigned true. If p is assigned false, the first clause can only be satisfied if q is assigned true. Therefore, either r must be assigned true or q must be assigned true, which is exactly what the new clause $\{q, r\}$ expresses. This idea can be generalized to clauses of the form $\{p, q_1, q_2, \ldots, q_n\}$ and $\{\neg p, r_1, r_2, \ldots, r_m\}$ with $n \geq 0$ and $m \geq 0$. These two clauses can be replaced by the single clause $\{q_1, q_2, \ldots, q_n, r_1, r_2, \ldots, r_m\}$.

The classic logical inference rule is *modus ponens*: p and $p \to q$ imply q. The two formulae are expressed by the clauses $\{p\}$ and $\{\neg p, q\}$. We can apply resolution to replace these two clauses by $\{q\}$. Resolution, therefore, is simply a generalized modus ponens.

If the clauses are of general form (not Horn clauses), the problem of satisfiability becomes much more difficult to solve. This is the reason PROLOG rules are restricted to Horn form. The implication $p_1 \wedge p_2 \wedge \ldots \wedge p_n \to q$ is just the PROLOG rule

$$q \text{ :- } p_1, p_2, \ldots, p_n.$$

2.5 Deduction

The syntax (:-) expresses the fact that the literals on the right side are negated. A fact has one single positive literal and is also a Horn clause.

If the predicates can contain variables, the simple equality test to look for pairs of p, $\neg p$ is replaced by unification. We have seen that unification makes two unified terms equal. Resolution can then be used. The unifying variable bindings have to be applied to the remaining predicates q and r as well. Recall that this is exactly what happens inside the query evaluator after unifying the left side of a rule with the goal.

The classical example with unification is the following.

This implication says that men are fallible.	`In[1]:= Assert[ fallible[x_], man[x_] ]`
Socrates is a man.	`In[2]:= Assert[ man[Socrates] ]`
Therefore, he is fallible.	`In[3]:= Query[ fallible[Socrates] ]` `Out[3]= Yes`

2.5.2 Deductive Databases and Expert Systems

Deductive databases are a combination of databases and logic inference. Databases enter naturally into logic programming because the collection of logic facts can be viewed as a database. The tables and relationships of databases can be realized easily as logic predicates (see Chapter 5 of the first volume for an explanation of basic concepts of databases). Here is a sample database about dinosaurs and humans, inspired by a well-known movie, based on a novel by Michael Crichton [13]. It lists a few of the characters appearing.

```
Assert[ dinosaur[Tyrannosaur] ]
Assert[ dinosaur[Velociraptor] ]
Assert[ dinosaur[Brontosaur] ]

Assert[ human[Grant] ]
Assert[ human[Tim] ]
Assert[ human[Lex] ]
```

Properties (attributes) of these entities can be defined by additional facts. Because the movie is mainly about who eats whom, we are interested in the composition and dietary habits of the main characters and their food.

```
Assert[ carnivore[Tyrannosaur] ]
Assert[ carnivore[Velociraptor] ]
Assert[ carnivore[Tim] ]

Assert[ vegetarian[Brontosaur] ]
Assert[ vegetarian[Lex] ]
```

```
Assert[ veryBig[Tyrannosaur] ]
Assert[ veryBig[Brontosaur] ]
Assert[ plant[Tree] ]
Assert[ plant[Grass] ]
Assert[ meat[cow] ]
Assert[ meat[goat] ]
```

The idea behind deductive databases is that not all facts need to be entered case by case. Some facts can be deduced logically from others, as we can do in PROLOG. For example, the cow and the goat are not the only characters consisting of meat. There is no need to enter each of the others explicitly, because we know that all dinosaurs and humans are in this class. The following deductive rules replace a larger number of individual assertions:

```
Assert[ meat[x_], dinosaur[x_] ]
Assert[ meat[x_], human[x_] ]
```

Relations between entities can be entered as predicates with several arguments. The relationship "x can eat y" is implemented by a logic predicate eat[x, y]. Carnivores prefer meat, unless it's too big to be eaten (or to be caught in the first place), and vegetarians prefer plants:

```
Assert[ eat[x_, y_], carnivore[x_], meat[y_], !veryBig[y_] ]
Assert[ eat[x_, y_], vegetarian[x_], plant[y_] ]
```

All facts and rules given above are part of the package JurassicPark.m.

Database queries take the form of ordinary logic queries. One question raised in the scene in which Dr. Grant, Tim, and Lex are high up in a tree staring at a Brontosaur, is whether the Brontosaur would eat Lex, because she is—as Tim jokingly points out—also a vegetarian, just like the animal confronting them.

Reassuringly, she is safe.	In[1]:= Query[eat[Brontosaur, Lex]] Out[1]= No
The Velociraptor enjoys a varied diet.	In[2]:= QueryAll[eat[Velociraptor, y_]] {y -> cow} {y -> goat} {y -> Velociraptor} {y -> Grant} {y -> Tim} {y -> Lex}

Deductive databases add such inference capabilities to ordinary database operations. One important class of rules is consistency constraints, which can be used to check that the database is consistent.

This consistency check ensures that we did not erroneously enter an item as both a dinosaur and a human.	`In[3]:= Query[ dinosaur[x_] && human[x_] ]` `Out[3]= No`

This query checks that all dinosaurs have been classified as either carnivore or vegetarian, that is, that our information is complete.	`In[4]:= Query[ dinosaur[x_],`
	`              !(carnivore[x_] || vegetarian[x_]) ]`
	`Out[4]= No`

Expert systems emphasize the deductive part of the database even more. Additional features of expert systems include an *explanation component* that tells the user how an answer was derived, and *fuzzy reasoning* that can deal with incomplete or contradictory information and with inferences that have probabilities attached to them. Such features have been used in medical expert systems, for example, where it is quite common that symptoms observed in a patient can be explained by different causes. The *expert system shell* takes the place of our simple query evaluator and offers these additional capabilities.

The preparation of the knowledge base is the crucial step in developing an expert system. The expert who provides the knowledge is often aided by a knowledge engineer, a specialist who knows how to put the expert's knowledge into the right form. Most expert systems have not developed out of PROLOG, but have their roots in artificial intelligence. They are, to date, the most successful application of this discipline, which has otherwise not fulfilled the promises with which it was advertised in the seventies and the eighties.

PROLOG has been used successfully in many commercial software projects, especially in *configuration programs*. The configuration of a system, for example, a computer system, is characterized by a database of available components and by rules about the interdependence of such components. Certain components require others (for example, a printer requires a printer port and a cable), while some components are mutually exclusive. Configuration programs let us check proposed configurations (such as a customer's order), or suggest configurations satisfying certain constraints (such as cost constraints).

2.6 Conclusions

The classical reference to PROLOG is the manual of one of its first implementations by Clocksin and Mellish [10]. Many examples in this chapter were taken from [8], which is an excellent introduction to PROLOG with many motivating examples from those parts of artificial intelligence that have turned out to be practical and useful.

PROLOG has always been regarded as something of a curiosity, an interesting idea but not for the real world. However, it has been used in a number of large industrial applications. In some cases, the development costs of a new PROLOG program were found to be less than the yearly maintenance of the old program written in the traditional way.

A database of real-world PROLOG applications is maintained by Prolog 1000, P.O. Box 137, Blackpool, Lancashire, FY2 0XY, U.K. Its description is as follows:

> The Prolog 1000 is a database of real Prolog applications being assembled in conjunction with the Association for Logic Programming (ALP) and PVG. The aim is to demonstrate how Prolog is being used in the real world and it already contains over 500 programs with well over 2 million lines of code. The database is available for research use in SGML format from the Imperial College archive . . . .

 The CD-ROM contains the packages LogicProgramming.m, Unify.m, DAG.m, Lisp.m, FSA.m, NIM.m, Lists.m, and JurassicPark.m, as well as the notebook LogicExamples.nb, containing the examples from this chapter.

Chapter Three

Higher-Order Functions

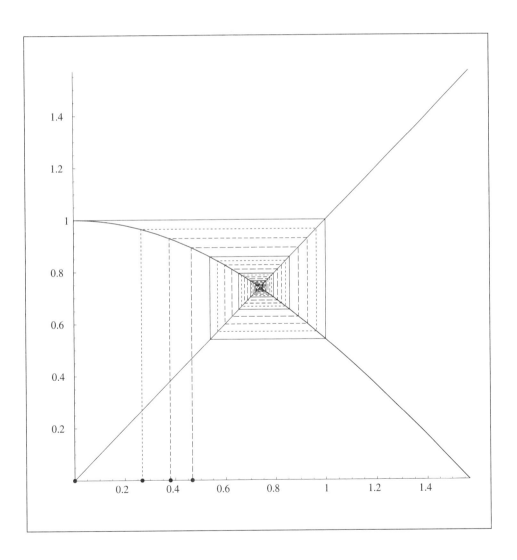

Although functions are arguably the most important concept in mathematics, functions are not generally available as (first-class) values in all programming languages. The exceptions are, of course, the languages that derive their collective name from functions: the functional languages. After a short introduction to functional languages, this chapter concentrates on higher-order functions: functions that take other functions as arguments or return functions as results. This possibility allows for a number of fascinating applications, ranging from useful to esoteric.

After an introduction to functional programming, Section 2 discusses the functional features of *Mathematica*. Next, we treat higher-order functions and their applications, such as mapping and folding, and we show how procedural control structures can be replaced by functional code. The last section is about a more theoretical aspect of functions: fixed-point operators and the theory of complete partial orders. These concepts can be realized in *Mathematica* and we can give simple examples of their usefulness.

The initial effort required to get acquainted with the notion of functions as values is well worth it. Thinking functionally, we can solve a range of problems far more elegantly than with the usual loop, which we often use first, without thinking, due to the lopsided introduction to programming to which most of us were subjected. *Mathematica* offers both worlds and is an ideal vehicle for experiments in programming style. This chapter shows some of the not-so-obvious uses of higher-order functions. It shows that programs themselves are values that can be manipulated by other programs.

About the illustration overleaf:

The fixed-point iteration for the cosine. Shown are four initial values and their first 10 iterates under the cosine function. The procedure `FunctionIteration[]` is developed in Chapter 7.

The code for this picture is in BookPictures.m.

3.1 Introduction

A functional programming language is a language that treats functions as "first-class" objects. This means that you can work with functions in the same way that you can work with values of other types (integers, for example) in ordinary languages. Among the possible options are the following:

- Create values of functional type at run time.
- Pass functions as arguments to other functions.
- Return functions as values of a function call.
- Include functions in data structures such as lists, arrays, or records.

The earliest functional language is LISP [42] (a particularly clean implementation is SCHEME [1, 56]). Modern functional languages include HASKELL [31], HOPE [3], MIRANDA [58], and ML [27]. The trend in functional language design is toward typed languages and lazy evaluation. Because *Mathematica* is an untyped language that uses strict evaluation, we shall not discuss these issues further. Pointers to information about functional programming are listed on the CD-ROM.

Traditional procedural languages may provide some of the features of functional languages, but the compilers of these languages severely restrict their ability to construct functions at run time. C and PASCAL allow you to pass functions as arguments. A function in C may even return a function as result. This capability is absent from block-structured languages (such as PASCAL) that allow the declaration of local functions within other functions. It is impossible to return a locally defined function as a result, because the stack frame in which it was defined ceases to exist when a function returns. Any references to local variables in the returned function become invalid at that point. (The stack frame is a piece of memory allocated on the stack; it is used to hold the local variables of a function. When the function returns to its caller, the stack frame is released.) Functional languages use more advanced models of variable binding, which can cope with such situations. Object-oriented languages allow functions (usually called methods) to be part of data structures (objects); see also Section 1.3.6.

3.2 The Functional Features of *Mathematica*

Mathematica provides all features of an untyped functional language. As we have seen, the main characteristic that distinguishes functional languages from the rest of the crowd is the ability to create new functions at run time. In *Mathematica* a function is created by an expression of the form Function[*var*, *body*], where *var* is a symbol and *body* is any expression, possibly containing *var*. The construct Function[*var, body*] is a function with formal parameter *var*. The function can be applied to an argument *arg* in the form Function[*var, body*][*arg*]. The function value is obtained by replacing every occurrence of *var* in *body* by *arg*, essentially by performing the substitution *body* /. *var* -> *arg* (there are some subtleties with the scoping of variables; see [40] for more explanations). The notation is borrowed from mathematics, where the function that squares its argument, for example, is often written in the form

$$f: x \mapsto x^2$$

This notation is translated into *Mathematica* to become f = Function[x, x^2].

This expression represents the function that squares its argument.

```
In[1]:= sq = Function[ x, x^2 ]
Out[1]= Function[x, x^2]
```

It can be applied to an argument.

```
In[2]:= sq[ 4 ]
Out[2]= 16
```

It can be differentiated. The result is the first derivative, another function.

```
In[3]:= sq'
Out[3]= Function[x, 2 x]
```

Most functional languages use a variant of lambda abstraction to define functions. In LISP you would write (lambda (x) (* 2 x)) to define the function that doubles its argument. In *Mathematica*, this becomes Function[x, 2 x], as we have seen. In ML, the definition looks like fn x => 2*x. The common characteristic is the need for a variable to denote the formal parameter of the function. Although the name of this variable does not matter, the fact that a name has to be chosen can lead to problems. One consequence is that formal parameters need to be renamed sometimes. This renaming is the source of the variables such as x$ that show up in *Mathematica* even if you never used them yourself.

Here is a definition of a function that returns another function as the result.

```
In[4]:= add[y_] := Function[x, x + y]
```

When the definition is evaluated the formal parameter x is renamed x$. In this case, the renaming was not necessary, but it was performed anyway for consistency.

```
In[5]:= add[ y ]
Out[5]= Function[x$, x$ + y]
```

Here, however, renaming is necessary: This function adds x to its argument, as we expect.

```
In[6]:= add[ x ]
Out[6]= Function[x$, x$ + x]
```

If no renaming had taken place, we would have gotten this function; it doubles its argument.

```
In[7]:= Function[x, x + x]
Out[7]= Function[x, x + x]
```

3.3 Functions as Data

There are other functional constructs in which an expression and a variable are given. Examples are `Plot[`*expr*`, {`*x*`, ...}]`, which plots the function `Function[`*x*`,` *expr*`]`; `D[`*expr*`,` *x*`]`, which forms the derivative of `Function[`*x*`,` *expr*`]`; and `Integrate[`*expr*`,` *x*`]`, which finds the antiderivative. Especially in the latter two cases, the fact that the variable is only loosely attached to the expression could be criticized by purists. The result returned does not even tell us what the variable was.

If we differentiate the function $x \mapsto xy$, the resulting expression does not make it clear that it is a function of x.	`In[8]:= D[x y, x]` `Out[8]= y`
The functional form is clearer; it leaves no doubt about the fact that the result is a function of x.	`In[9]:= Function[x, x y]'` `Out[9]= Function[x, 1 y]`

3.3 Functions as Data

Functions that take other functions as arguments or return functions as results are often called *higher-order functions*. The derivative operator `'` (`Derivative[1]` in `FullForm`) is a simple example of a higher-order function.

`Derivative[1]` can be applied to a function and it returns another function as result.	`In[1]:= Derivative[1][ Sin ]` `Out[1]= Cos[#1] &`

3.3.1 Mapping and Folding

Among the simplest higher-order functions are the mapping and folding operations that apply a function passed as an argument in various ways to data structures.

The simplest one is `Map`. It applies its first argument to all elements of the second argument and returns the resulting structure.	`In[2]:= Map[ f, {a, b, c} ]` `Out[2]= {f[a], f[b], f[c]}`
We can either use an existing (built-in or user-defined) function as the first argument of `Map`,	`In[3]:= Map[ Sqrt, {1, 2, 3, 4} ]` `Out[3]= {1, Sqrt[2], Sqrt[3], 2}`
or we can use a pure function, created on the fly.	`In[4]:= Map[ Function[x, x^2], % ]` `Out[4]= {1, 2, 3, 4}`

Nest applies a function repeatedly to the result of the previous application, starting with x.

```
In[5]:= Nest[ f, x, 5 ]
Out[5]= f[f[f[f[f[x]]]]]
```

This iteration finds an approximation of the Golden Ratio, the solution of the fixed-point equation $x = 1 + 1/x$.

```
In[6]:= Nest[ Function[x, 1 + 1/x], 1.0, 10 ]
Out[6]= 1.61798
```

FixedPoint[] takes a function as its argument and returns its fixed point. The fixed point of this function is $\sqrt{2}$.

```
In[7]:= FixedPoint[ Function[x, (x + 2/x)/2], 2.0 ]
Out[7]= 1.41421
```

Fold takes a function of two arguments as its first argument and repeatedly applies it to the previous result and an element from the list in turn, starting with x0.

```
In[8]:= Fold[ g, x0, {x1, x2, x3} ]
Out[8]= g[g[g[x0, x1], x2], x3]
```

There is no guarantee that functional code is always clear and readable. This example constructs the set of all subsets of {a, b, c}.

```
In[9]:= Fold[ Function[{sets, elem},
              Function[s, Sequence @@ {s, Append[s, elem]}] /@
              sets ], {{}}, {a, b, c} ]
Out[9]= {{}, {c}, {b}, {b, c}, {a}, {a, c}, {a, b},
        {a, b, c}}
```

3.3.2 Control Structures

The functional operations just mentioned embody some of the control structures of conventional languages. This is the reason that control structures such as loops are not necessary in functional languages. They can be replaced by recursion or functional operations.

Loops over the elements of a list (or an array) can usually be replaced by mapping. The functional version of a program to square all elements of a list, for example, uses mapping to perform the required operation on all elements.

```
squareList[l_] := Map[ Function[e, e^2], l ]
```

The procedural version requires an explicit loop and a few auxiliary variables. We assumed here that lists are used in the same way that arrays are used in procedural languages, that is, we access their elements with $l[[i]]$.

```
squareListLoop[l_] :=
    Module[{n = Length[l], r, i},
        r = Table[0, {n}];
        Do[ r[[i]] = l[[i]]^2, {i, 1, n} ];
        r
    ]
```

3.3 Functions as Data

Simple recursive definitions can be written in terms of `Fold`. The operation to fold over is the function that combines the current value of the recursion parameter with the result of the recursive call. Let us look at the standard example, the factorial function. Its recursive definition is

```
fact[0]  =  1
fact[n_] := n * fact[n-1]
```

The binary operation combining n with the recursive result is simply multiplication. The recursion parameter ranges from 1 to n. The initial value is 1; it is obtained from the first rule. This gives the following definition:

```
factFold[n_] := Fold[ Times, 1, Range[n] ]
```

A loop that reduces a list to a single value is usually expressed recursively in functional languages. Here is a procedural program that computes the sum of the elements of a list using an auxiliary variable in a loop.

```
sumIterative[l_List] :=
    Module[{sum},
        sum = 0;
        Do[ sum += l[[i]], {i, 1, Length[l]} ];
        sum
    ]
```

The recursion happens along the tail of the list. (Many functional languages represent lists in the same way as does LISP. The primitive operations are `First`, `Rest`, `Prepend`, and the empty list `{}`.)

```
sumRecursive[{}] = 0
sumRecursive[{first_, rest___}] := first + sumRecursive[{rest}]
```

This recursion, too, can be expressed using `Fold`. The initial value is 0, obtained from the first rule. The binary operation is addition, and the recursion parameter ranges over the whole list:

```
sumFold[l_] := Fold[ Plus, 0, l ]
```

The transformation of programs from recursive into iterative or functional form is an important tool for optimizing compilers. Program transformations (often in the reverse direction) can also be used to prove the correctness of programs.

3.3.3 Currying

Some functional languages restrict functions to take only one argument. It is still possible to write functions that take more than one argument, by nesting function calls. Instead of g[x, y] one uses cg[x][y], where cg is the so-called *Curried* version of g. (The term is named for the logician H. B. Curry.) The definition of cg is easy.

This definition sets up cg as a function of one argument.	In[10]:= cg[x_] := Function[y, g[x, y]]
The Curried form of g is equivalent to the ordinary form.	In[11]:= cg[a][b] Out[11]= g[a, b]

Because the arguments of a Curried function are supplied one at a time in the form cg[x][y], there is an intermediate expression cg[x]. Its value is a pure function.

The intermediate expression cg[a] is a pure function in which the value of the argument a has been frozen for later use when the second argument is supplied.	In[12]:= cg[a] Out[12]= Function[y$, g[a, y$]]
Now we are back at g[a, b].	In[13]:= %[b] Out[13]= g[a, b]

Note that there are essentially three different ways to define cg:

```
cg = Function[x, Function[y, g[x, y]]]
cg[x_] = Function[y, g[x, y]]
cg[x_][y_] = g[x, y]
```

As an exercise, you may want to think about the differences among these three forms.

The Curried version of a binary operator is a higher-order function that returns a function as the result, as we saw in the preceding example. We can easily develop a function that constructs Curried functions. BinaryCurry[g] returns the Curried version of the binary operator g.

```
BinaryCurry[ g_ ] := Function[x, Function[y, g[x, y]]]
```

Here is the Curried version of the addition operator Plus.	In[14]:= cPlus = BinaryCurry[Plus] Out[14]= Function[x$, Function[y$, x$ + y$]]
If we give it only its first argument, we get the function inc that adds 1 to its argument.	In[15]:= inc = cPlus[1] Out[15]= Function[y$, 1 + y$]
Here is an example of its use.	In[16]:= inc[5] Out[16]= 6

3.4 Fixed Points of Higher-Order Functions

There seems to be one fundamental limitation of pure (anonymous) functions. Because a pure function does not have a name, it seems impossible to write recursively defined pure functions. Let us have a look at the standard example: the factorial function. If we give it a name, we can easily define it recursively.

```
fact[n_] := If[ n == 0, 1, n fact[n-1] ]
```

But how would we write a pure function that performs the same computation? It is indeed impossible to write simple pure functions in LISP or ML that perform these tasks. *Mathematica* offers a back door. The pure function itself is available as "parameter" zero, written #0; therefore, we can write

```
Function[ If[#1 == 0, 1, #1 #0[#1-1]] ]
```

The pure function Function[#0], or #0&, shows how #0 works. When the function is applied, #0 is replaced by the head of the expression containing it, that is, the pure function itself.

```
In[17]:= Function[#0][]
Out[17]= #0 &
```

This feature is also the source of the shortest *Mathematica* program known so far that *prints* itself.

This solution was independently found by Michael Trott and the author; the problem was formulated by Todd Gayley for the *Mathematica* user group mailing list.

```
In[18]:= Print[ToString[#0][]] & []
Print[ToString[#0][]] & []
```

Here we compute 10! recursively, using a pure function.

```
In[19]:= Function[ If[#1 == 0, 1, #1 #0[#1-1]] ] [ 10 ]
Out[19]= 3628800
```

The problem of naming pure functions is not as artificial as it may seem. The issue can be quite subtle, as this example shows.

We can name a pure function and use its name inside the function, as in this definition of the factorial function.

```
In[20]:= gfact = Function[n, If[n==0, 1, n gfact[n-1]]];
```

```
In[21]:= gfact[10]
Out[21]= 3628800
```

The same definition, used as a local constant in a `With`, does not work. Can you see what is going on?

```
In[22]:= With[{lfact =
              Function[n, If[n==0, 1, n lfact[n-1]]]},
          lfact[10]
         ]
Out[22]= 10 lfact[9]
```

A more general and cleaner approach allows the definition of any programmable function as a pure function. This approach works in most functional languages. The required tool is the fixed-point or **Y** combinator, introduced by Curry.

3.4.1 Fixed-Point Iteration

Fixed points and methods to compute them occur in many different settings. In general, we require a domain of objects or points and a set of functions that map the domain into itself (that is, the functions take an element of the domain as the argument and give an element as their result).

One of the simplest examples of such a domain is the set of real numbers and the real-valued functions. Such functions may have fixed points. The cosine, for example, has a fixed point near $x_c = 0.739085$, that is, $\cos x_c = x_c$. Fixed points can often be found by iteration.

The computation that starts with the value 1.0 and applies the cosine repeatedly converges to the fixed point.

```
In[23]:= Nest[ Cos, 1.0, 100 ]
Out[23]= 0.739085
```

The `FixedPoint` operator performs as many iterations as are necessary to reach the fixed point.

```
In[24]:= FixedPoint[ Cos, 1.0 ]
Out[24]= 0.739085
```

Note, however, that the exact fixed point cannot be found by a finite iteration. The exact fixed point can be expressed only by the infinite formula $\cos(\cos(\ldots(1)\ldots))$.

Let us now turn to a more abstract domain, where we can hope to construct *functions* as fixed points. The elements of our domain are programs. To be more precise, they are functions over the integers, defined as *Mathematica* programs. A simple example is the identity function, written `Function[n, n]`. Our functions need not be defined for all possible argument values. The square root, for example, is defined only for integers that are perfect squares (remember that we work only over the integers). Another important function, called `bottom`, is not defined anywhere. It can be implemented in this way, among others.

`bottom = Function[n, While[True, 0]]`

The loop in the body of the function never terminates; therefore, this function returns no value, regardless of its argument.

3.4 Fixed Points of Higher-Order Functions

Now that we have defined our domain of objects, let us look at the mappings from the domain of *Mathematica* functions into itself. These are our higher-order functions. They take a function (of integers) as their argument and return another such function. We shall call such a higher-order function a *functional*, to avoid confusion with the ordinary functions on which it operates. Here is an example of a functional.

```
mFact = Function[ f, Function[ n, If[n == 0, 1, n f[n-1]] ] ]
```

The functional mFact takes a function f as argument and returns another function, given as a pure function. To see what it does, let us apply it to various functions and examine the results returned.

The result of applying mFact to bottom is a function f1 that gives 1 for $n = 0$ and is undefined for other arguments.

```
In[25]:= f1 = mFact[bottom]
Out[25]= Function[n$, If[n$ == 0, 1, n$ bottom[n$ - 1]]]
```

Here we apply mFact to the factorial function. The resulting function gives 1 for $n = 0$ and gives $n(n-1)!$ for arguments $n > 0$. Because $n(n-1)! = n!$ and $0! = 1$, this function is the factorial function. In other words, the factorial function is a fixed point of mFact.

```
In[26]:= mFact[ Factorial ]
Out[26]= Function[n$, If[n$ == 0, 1, n$ (n$ - 1)!]]
```

The fact that functions arise as fixed points of functionals can be seen in another way. Consider the recursive definition of the factorial function gfact, given earlier:

```
gfact = Function[n, If[n==0, 1, n gfact[n-1]]]
```

This definition can be viewed as an equation for the unknown function gfact. We can write the right side in the form

```
Function[g, Function[n, If[n==0, 1, n g[n-1]]][gfact]
```

The factorial function is, therefore, a solution of the fixed-point equation

```
gfact == F[gfact]
```

where

```
F = Function[g, Function[n, If[n==0, 1, n g[n-1]]] .
```

The domain in which this equation has a solution is the space of all *Mathematica* programs. We shall see that we can use iteration to find the solution.

It is straightforward to find a functional that has a desired recursively defined function g as its fixed point. Start with the definition of g, assumed to consist of a single rule of the form

$$g[n\_] := body$$

where the expression *body* usually contains at least one conditional and may contain recursive occurrences of g. Write this definition as a pure function:

$$g = Function[\ n,\ body\ ]$$

Then, turn the pure function into a definition for a functional whose argument is g:

$$makeG[g\_] := Function[\ n,\ body\ ]$$

If you wish, you can also write it explicitly as a pure function:

$$makeG = Function[\ g,\ Function[n,\ body]\ ]$$

This is the form we used for mFact above.

The remaining question is how to find the fixed points of functionals in a systematic way. It turns out that we can perform an iteration, quite similar to the iteration we used to find the fixed point of the cosine. We already began the iteration for mFact above. We used bottom as the initial value to start the iteration. We saw that f1 = mFact[bottom] agrees with the factorial function for argument 0 and is undefined otherwise. Let us continue this iteration for a few more steps.

After two iterations of mFact we get a function that agrees with the factorial for arguments 0 and 1 and is undefined otherwise.

```
In[27]:= f2 = mFact[f1]
Out[27]= Function[n$, If[n$ == 0, 1,
    n$ Function[n$, If[n$ == 0, 1, n$ bottom[n$ - 1]]][
    n$ - 1]]]
```

Ten iterations give a complicated mess that agrees with the factorial for arguments 0, 1, ..., 9 and is undefined otherwise.

```
In[28]:= Short[ f10 = Nest[ mFact, bottom, 10], 5 ]
Out[28]//Short=
  Function[n$, If[n$ == 0, 1,
    n$ Function[n$, If[n$ == 0, 1,
      n$ Function[n$, If[n$ == 0, 1,
        n$ Function[n$, <<1>>][n$ - 1]]][n$ - 1]]][
    n$ - 1]]]
```

Here we see that it indeed gives us 9!. The computation f10[10] would not terminate, however.

```
In[29]:= f10[ 9 ]
Out[29]= 362880
```

3.4 Fixed Points of Higher-Order Functions

By iterating `func` with initial value `bottom` we can find better and better approximations of the fixed point, the factorial function. Let us present some of the theory on which this construction is based: complete partial orders.

3.4.2 Complete Partial Orders

The notion of approximation of functions is an important tool in the field of semantics of programming languages. A function f approximates g, written $f \sqsubseteq g$, if the domain of definition of f is contained in the domain of g (that is, $\text{dom } f \subseteq \text{dom } g$), and if the two functions agree on their common domain (which is the domain of f), that is, $g(x) = f(x)$ for all $x \in \text{dom } f$. The relation $\sqsubseteq$ introduces a partial order on the space of all functions.

A relation $\sqsubseteq$ is a partial order if for any functions f, g, and h the following three properties are true:
$$f \sqsubseteq f$$
$$f \sqsubseteq g \text{ and } g \sqsubseteq f \text{ implies } f = g \quad (3.4\text{--}1)$$
$$f \sqsubseteq g \text{ and } g \sqsubseteq h \text{ implies } f \sqsubseteq h.$$

The nowhere-defined function, which we called `bottom`, is denoted $\bot$. It is the least element in the space of all functions, that is, $\bot \sqsubseteq f$ for all f. Our functionals are continuous functions on this space. An important theorem states that all continuous functions on complete partial orders have a fixed point.

A partial order is *complete* if all completely ordered subsets (called chains) C have a *least upper bound* (lub), denoted $\bigsqcup C$. A functional F is *monotone* if $f \sqsubseteq g$ implies $F(f) \sqsubseteq F(g)$. A monotone functional is *continuous* if $F(\bigsqcup C) = \bigsqcup F(C)$ for all chains C.

The (least) fixed point of the functional F can be found by
$$\bigsqcup_{k=0}^{\infty} F^k(\bot). \quad (3.4\text{--}2)$$

(F^k denotes k-fold iteration of F.)

In our example, F was `mFact`, and $\bot$ was `bottom`. We performed only a finite number of iterations, and as a consequence got a result that approximates the fixed point. If we could perform infinitely many iterations, we could construct the fixed point exactly.

Fortunately, we do not need the exact fixed point to compute factorials. For any given argument n, a finite approximation with at least $n + 1$ iterations suffices, as we have seen above where we computed 9! using `f10`. The definition of approximation ensures that if an approximation is defined for a certain argument, its result agrees with the result of the approximated function.

No *a priori* number of iterations can be set, because we must be prepared for an argument n of any size. We can, however, write a program that performs enough iterations for any given argument and then calls this approximation function:

```
factNested[n_] := Nest[mFact, bottom, n+1][n]
```

Because the names `mFact` and `bottom` used here are merely shorthand for pure functions, we have managed to give a recursive definition of the factorial function without giving it a name.

The function `factNested` can be applied to any argument. It *is* the factorial function, admittedly written in a rather complicated way.

```
In[30]:= factNested[100] // N
Out[30]= 9.33262 10^157
```

Note that the undefined function `bottom` is not really needed as an initial value. Anything could be inserted in its place, for example, `Null` or even `mFact` itself. (A similar observation is true for the iteration that finds the fixed point of the cosine: Any real number within the range of convergence of the iteration can be used.)

For functionals derived from recursive definitions in the way shown above, the least fixed point will be the desired function. For other functionals the fixed point may well be `bottom` itself, that is, the completely undefined function. A trivial example of such a trivial functional is the constant `Function[f, bottom]`.

3.4.3 The Y Combinator

Our construction to iterate a functional as often as needed, used in the definition of `factNested`, can be generalized. There is an amazing higher-order function Y that takes any functional as an argument and returns its fixed point as a pure function.

```
Y[F_] := With[{g = Function[f, Function[x, F[f[f]][x] ]]}, g[g] ]
```

Y.m: the applicative-order fixed-point combinator

To understand how the **Y** combinator works, observe that it constructs a program that performs another step in the fixed-point iteration whenever there is a recursive call. The main idea is to pass itself as an argument so that the code remains available for further iterations, whenever needed. Because the initial value of the fixed-point iteration is irrelevant, as we have seen, it is a convenient place to pass along the whole code. If the recursion is well formed, it will stop eventually and the code for the base case terminates the iteration. A step-by-step construction of **Y** in SCHEME is shown in the delightful book [20].

3.4 Fixed Points of Higher-Order Functions

The **Y** combinator constructs the factorial function as the fixed point of the functional **mFact**. The result is rather incomprehensible.

```
In[31]:= factY = Y[mFact]
Out[31]= Function[x$, Function[f,
            Function[n, If[n == 0, 1, n f[n - 1]]]][Function[f$,
            Function[x$, Function[f,
               Function[n, If[n == 0, 1, n f[n - 1]]]][f$[f$]][
               x$]]][Function[f$,
            Function[x$, Function[f,
               Function[n, If[n == 0, 1, n f[n - 1]]]][f$[f$]][
               x$]]]][x$]]
```

But the code works for any argument (provided the recursion limit $RecursionLimit is set high enough).

```
In[32]:= factY[200] // N
Out[32]= 7.88657867364791 10^374
```

For another example, here is the functional whose fixed point is the Fibonacci function.

```
In[33]:= makeFib[f_] :=
            Function[ n, If[n<=2, 1, f[n-1] + f[n-2]] ]
```

We construct its fixed point.

```
In[34]:= fib = Y[makeFib];
```

Here is a table of the first ten Fibonacci numbers.

```
In[35]:= Table[fib[i], {i, 10}]
Out[35]= {1, 1, 2, 3, 5, 8, 13, 21, 34, 55}
```

You may ask yourself whether a similar construction could find the fixed point of the cosine, the example with which we started. There is a simpler, more general fixed-point combinator that works for arbitrary domains, given by

```
Y1[F_] := With[{g = Function[ f, F[f[f]] ]}, g[g] ]
```

but it is of no use in strict functional languages, such as *Mathematica*. The self application `f[f]` always leads to an infinite evaluation chain. The reason this does not happen with **Y** is that the self application occurs only inside the body of a pure function. The body of a function is not evaluated when the function is defined, only later when and if it is actually used. Because only a finite approximation of the fixed point is needed, the self application goes away eventually. For (numerical) fixed points of functions on the real numbers, however, no finite approximation gives us the true fixed point.

Readers interested in the mathematics of fixed points can find an account of the theory of complete partial orders and their application to the semantics of programming languages in the textbook [15].

 The CD-ROM contains the package Y.m, as well as the notebook FixedPoints.nb, containing the examples from this chapter.

Chapter Four

Combinators

Combinators are an alternative to pure functions. They do not use variables and are, therefore, immune to the scoping problems caused by conflicts of names of formal parameters. We present an introduction to combinatory algebras and show how to convert pure functions into combinators.

After an introduction, we discuss combinatory algebras in Section 2. They are the theoretical background of the constructs used later. In Section 3 we show how expressions can be turned into combinators, by a method called combinatory abstraction. Then, in Section 4, we use combinatory abstraction to convert pure functions into terms without variables. The implementation of these ideas requires control of the order of substitutions in rewrite rules. We discuss the techniques needed to achieve this control. A section on applications concludes the chapter. We resume the discussion of fixed points from Section 3.4 and present some of the theory of computation within combinatory algebras.

This chapter shows once more that *Mathematica* is well suited to implementing formal systems and to experimenting with them. Here, we use expressions that have deeply nested heads. We show how we can specify the order of evaluation using replacement rules with conditions.

About the illustration overleaf:

A sunflower. The points are arranged in a spiral fashion. The angle between two points is $\frac{2\pi}{\varphi}$, where φ is the Golden Ratio. The radius of the ith point is proportional to $\sqrt{i}$. The idea for the sunflower program came from [49].

The code for this picture is in BookPictures.m.

4.1 Introduction

Chapter 3 introduced pure functions, which look like Function[*x*, *body*] in *Mathematica*. As we saw, the fact that a name has to be chosen for the formal parameter can lead to problems.

Here is a variant of the Curried addition from Chapter 3.

```
In[1]:= add = Function[y, Function[x, x + y]]
Out[1]= Function[y, Function[x, x + y]]
```

When the definition is evaluated, the formal parameter x is renamed to x$.

```
In[2]:= add[y]
Out[2]= Function[x$, x$ + y]
```

Here we see why the renaming is necessary. This function adds x to its argument, as we expect. If no renaming had taken place, we would have gotten Function[x, x + x], a function that doubles its argument.

```
In[3]:= add[x]
Out[3]= Function[x$, x$ + x]
```

There is an alternative way to define functions. The main idea is not to use any variables at all to describe functions.

4.2 Combinatory Algebras

Functional languages modeled after λ-calculus provide two basic operations: application (of functions to arguments) and abstraction. Abstraction takes a term t and a variable x and forms the function $\lambda x.t$ or Function[x, t].

A combinatory algebra is a structure with just one operation: application. Because there is no way to define functions by abstraction, some basic operations need to be built in. These are the constants or basic combinators. Combinatory terms are built up from the constants and variables by repeated application. In the literature, the application of term t_1 to term t_2 is written $t_1 t_2$, omitting the operator (similar to the way multiplication is often written without the dot). Application is taken to be left associative; that is, xyz is interpreted as $(xy)z$, not as $x(yz)$.

Remarkably, two basic combinators suffice to express all possible functions. Let us see how this works. With abstraction we can turn any term t (involving the variable x) into a function Function[x, t]. If we want to define the same function as a combinatory term we must find a term T (not containing x) such that $Tx = t$. In this case, T describes the same function as Function[x, t]. The term t is either an atom (a variable or a constant) or it is composite,

$t = t_1 t_2$. For each of these possibilities, we can easily see what kind of combinators are necessary to find the corresponding T:

- t is the variable x. The term T must be the identity combinator **I**, with the property

$$\mathbf{I}X = X \qquad \forall X. \tag{4.2-1}$$

We have $Tx = \mathbf{I}x = x$, as required.

- t is a constant or variable c, other than x. The function Function[x, c] is constant, returning c no matter what its argument is. Therefore, we need a constructor for constant functions, the combinator **K** with the property

$$\mathbf{K}XY = X \qquad \forall X, Y. \tag{4.2-2}$$

We can set $T = \mathbf{K}c$. We have $Tx = \mathbf{K}cx = c$, as required.

- t is a composite term $t_1 t_2$. Recursively we can find combinatory terms T_1 and T_2 with the properties $T_1 x = t_1$ and $T_2 x = t_2$. To build T we need a combinator **S** that distributes an argument onto two terms:

$$\mathbf{S}XYZ = XZ(YZ) \qquad \forall X, Y, Z. \tag{4.2-3}$$

We can set $T = \mathbf{S}T_1 T_2$. We have $Tx = \mathbf{S}T_1 T_2 x = T_1 x(T_2 x) = t_1 t_2$, as required.

The combinator **I** is even redundant. We can use **SKK** instead, because $\mathbf{SKK}X = \mathbf{K}X(\mathbf{K}X) = X$, according to Equations 4.2–3 and 4.2–2. However, we shall keep working with all three combinators, to make our sometimes rather complicated terms more readable.

Formally, a *combinatory algebra* is a structure with one binary operation (application) that contains two elements **S** and **K** with the desired properties stated in Equations 4.2–3 and 4.2–2. Combinatory terms are defined inductively:

- A variable is a combinatory term.
- A constant is a combinatory term.
- If t_1 and t_2 are combinatory terms, then $t_1 t_2$ (the application of t_1 to t_2) is a combinatory term.

A *combinator* is a term that does not contain any variables. If the combinatory algebra does not contain any constants besides **S** and **K** (a pure combinatory algebra), combinators will be terms containing only **S** and **K**.

Combinatory algebras were developed by Schönfinkel and Curry in the 1920s [52, 14]. They studied systems of combinatory logic (logic without variables), including the **S** and **K** combinators. These systems were later found to be paradoxical and thus unsuitable for the description of logic. (We shall explain the paradoxical nature later, in Section 4.5.1.) Combinators have an obvious operational nature, however: They transform terms in various ways. As such it is no surprise that

4.2 Combinatory Algebras

interest in them was renewed in theoretical computer science. Programs, too, have an operational aspect; it seems natural to use λ-calculus or combinatory algebras to describe the semantics of programming languages.

In the formal treatment of combinatory logic, the combinators **S** and **K** define a reduction (term-rewriting) relation according to the rules

$$\begin{aligned} \mathbf{K}XY &\to X \\ \mathbf{S}XYZ &\to XZ(YZ) \end{aligned} \qquad (4.2\text{--}4)$$

A term that does not contain any reducible subterms is said to be in normal form. It is, of course, trivial to implement these rules in *Mathematica*.

More information about combinatory algebras can be found in the textbook [29] or the standard reference [5].

4.2.1 Implementation

In *Mathematica* we can use the familiar form of application of a function to an argument $t_1[t_2]$ to realize the operation of application $t_1 t_2$ in a combinatory algebra. This representation is also left associative: $t_1[t_2][t_3]$ stands for $t_1 t_2 t_3 = (t_1 t_2) t_3$, rather than for $t_1(t_2 t_3)$ (the latter is $t_1[t_2[t_3]]$). As a consequence, you will encounter expressions that are deeply nested in their head position, something that is rarely seen in other applications.

The symbols **S**, **K**, and **I** are so standard in the literature that we do not want to use different names. To protect them from interference with built-in objects (**I** being the main problem), we define them in their own context. The reduction rules for **S**, **K**, and **I** (Equation 4.2–4) are easily programmed. Listing 4.2–1 displays part of the package Combinators.m.

```
BeginPackage["Combinators`"]
`S::usage = "S is the distributive combinator."
`K::usage = "K is the constant combinator."
`I::usage = "I is the identity combinator."
Begin["`Private`"]
I[x_] := x
K[x_][y_] := x
S[x_][y_][z_] := x[z][y[z]]
End[]
Protect[S, K, I]
EndPackage[]
```

Listing 4.2–1: Part of Combinators.m

Note that we introduce the three symbols S, K, and I in the usage messages in the form `S, `K, and `I. The context marks (backquotes) force them to be created in the current context

Combinators`, even if they exist already in the system context. Without the initial context mark, I would refer to the existing symbol in the system context. Once defined, the symbol lookup rules (which search the system context last) guarantee that the unadorned form I refers to the combinator, rather than to the complex number.

The warning message alerts us to our nonstandard practice of declaring symbols that shadow system symbols.

```
In[4]:= Needs["MathProg`Combinators`"]
K::shdw: Symbol K appears in multiple contexts
    {MathProg`Combinators`, System`}; definitions in context
    MathProg`Combinators`
    may shadow or be shadowed by other definitions.

I::shdw: Symbol I appears in multiple contexts
    {MathProg`Combinators`, System`}; definitions in context
    MathProg`Combinators`
    may shadow or be shadowed by other definitions.
```

The **I** combinator hides the complex number I, which is what we want here.

```
In[5]:= ?I
I is the identity combinator.
```

The rule takes effect as intended.

```
In[6]:= I[x]
Out[6]= x
```

We can see that **I** could indeed be replaced by **SKK**. The two combinators behave in the same way.

```
In[7]:= S[K][K][x]
Out[7]= x
```

4.3 Combinatory Abstraction

In the previous section we gave the rules for converting a term t, involving the variable x, into a combinatory term T such that $Tx = t$. The operation of converting t to T is written $\lambda^* x.t$. According to Equations 4.2–1 through 4.2–3 it can be defined inductively as follows:

$$\begin{aligned} \lambda^* x.x &= \mathbf{I} \\ \lambda^* x.c &= \mathbf{K}c \quad \text{for atoms } c \neq x \\ \lambda^* x.t_1 t_2 &= \mathbf{S}(\lambda^* x.t_1)(\lambda^* x.t_2) \end{aligned} \quad (4.3\text{–}1)$$

These rules are sufficient, but they can give unnecessarily complicated results. We can make two obvious improvements.

First, we can treat composite terms not containing x as if they were constant atoms. If a constant term t is composed, $t = cd$, say, we get according to Equation 4.3–1

$$\lambda^* x.cd = \mathbf{S}(\lambda^* x.c)(\lambda^* x.d) = \mathbf{S}(\mathbf{K}c)(\mathbf{K}d) . \quad (4.3\text{–}2)$$

However, the simpler term $\mathbf{K}(cd)$ works as well, because $\mathbf{K}(cd)X = cd = S(\mathbf{K}c)(\mathbf{K}d)X$. We shall, therefore, use the rule $\lambda^* x.t = \mathbf{K}t$ for *all* terms t not containing x, not only for atoms.

4.3 Combinatory Abstraction

A second important special case is $\lambda^*x.t_1x$, where t_1 does not contain x. Instead of

$$\lambda^*x.t_1x = \mathbf{S}(\lambda^*x.t_1)(\lambda^*x.x) = \mathbf{S}(\mathbf{K}t_1)\mathbf{I} \qquad (4.3\text{--}3)$$

we can simply use t_1, because obviously $t_1x = t_1x$.

Additionally, we define a form of λ^* with several arguments according the the recursion

$$\lambda^*x_1x_2\ldots x_n.t = \lambda^*x_1.\lambda^*x_2.\ldots.\lambda^*x_n.t. \qquad (4.3\text{--}4)$$

The definitions for `LambdaStar[`x`][`t`]`, our implementation of $\lambda^*x.t$, are shown in Listing 4.3–1.

```
LambdaStar[x_Symbol][x_] := I
LambdaStar[x_Symbol][c_] /; FreeQ[c, x] := K[c]
LambdaStar[x_Symbol][f_[x_]] /; FreeQ[f, x] := f
LambdaStar[x_Symbol][a_[b_]] := S[LambdaStar[x][a]][LambdaStar[x][b]]
LambdaStar[x_Symbol, y__][t_] := LambdaStar[x][ LambdaStar[y][Unevaluated[t]] ]
```

Listing 4.3–1: Part of Combinators.m: Combinatory abstraction

The extra `Unevaluated` in the last rule preserves `t` in unevaluated form during the recursion, if `LambdaStar` was originally called as `LambdaStar[`x_1`, ..., `x_n`][Unevaluated[`t`]]`. (We shall need this form in the next section.) If no `Unevaluated` is present in the original call, nothing bad happens because `t` is evaluated in the original call and is then merely passed along.

These two expressions demonstrate the identity and constants, respectively.

```
In[8]:= {LambdaStar[x][x], LambdaStar[x][c]}
Out[8]= {I, K[c]}
```

We can always check the results by applying them to `x`. We should get the original expression back. (`Through[{`f_1`, `f_2`}[`$args$`]]` gives `{`f_1`[`$args$`], `f_2`[`$args$`]}`.)

```
In[9]:= Through[ %[x] ]
Out[9]= {x, c}
```

Here is the treatment of a composite constant.

```
In[10]:= LambdaStar[x][c d]
Out[10]= K[c d]
```

This result is the equivalent of the η rule in λ-calculus, which says that `Function[`x`, f[`x`]]` is the same as simply `f`.

```
In[11]:= LambdaStar[x][f[x]]
Out[11]= f
```

Here is an example where the function itself depends on `x`.

```
In[12]:= LambdaStar[x][x[e]]
Out[12]= S[I][K[e]]
```

As usual we can easily check the result.

```
In[13]:= %[x]
Out[13]= x[e]
```

Self-application leads to this term.	`In[14]:= LambdaStar[x][x[x]]` `Out[14]= S[I][I]`

When this term is applied to itself we get an infinite iteration. The term **SII(SII)** does not have a normal form. It has a single infinite reduction: **SII(SII)** → **I(SII)(SII)** → **SII(SII)**.	`In[15]:= %[%]`
	`$IterationLimit::itlim: Iteration limit of 4096 exceeded.`
	`Out[15]= Hold[I[S[I][I]][I[S[I][I]]]]`

Here is an example with two variables.	`In[16]:= LambdaStar[x, y][y[x][y]]` `Out[16]= S[S[K[S]][S[K[S[I]]][K]]][K[I]]`

4.3.1 Interfacing with Built-In Functions

Unary functions in *Mathematica* can simply be treated as constants in our combinatory algebra. The treatment of binary functions is not as straightforward. In combinatory algebras, functions have only one argument, as they do in λ-calculus. Functions of several variables can be Curried: Instead of `f[x, y]` we use `f[x][y]`. To interface with binary functions built into *Mathematica*, we can simply define a combinator that represents the Curried form of a function. We call it **c**. It has the property that

$$\mathbf{c}FXY = F(X,Y) \qquad \forall F,X,Y. \tag{4.3-5}$$

The corresponding rule for λ^* is

$$\lambda^* x.f(a,b) = \lambda^* x.\mathbf{c}fab. \tag{4.3-6}$$

We cannot implement this rule naively. The term $\mathbf{c}fab$ in the right side would turn back into $f(x,y)$ by Equation 4.3–5. Therefore, we apply the rule for **S** once to get $\mathbf{S}(\lambda^* x.\mathbf{c}fa)(\lambda^* x.b)$. Here are the necessary additions to our package:

```
`c::usage = "c[f] represents the binary function f in Curried form."
c[f_][x_][y_] := f[x, y]
LambdaStar[x_][f_[a_, b_]] := S[LambdaStar[x][c[f][a]]][LambdaStar[x][b]]
```

Here is the term representing a binary sum as a combinatory term.	`In[17]:= LambdaStar[x, y][x+y]` `Out[17]= S[S[K[S]][S[K[K]][c[Plus]]]][K[I]]`
It checks out as expected.	`In[18]:= %[a][b]` `Out[18]= a + b`
Abstracting over y first gives a different term. We have not expressed the fact that addition is commutative.	`In[19]:= LambdaStar[y, x][x+y]` `Out[19]= S[K[S[c[Plus]]]][K]`

Here is a more complicated example involving one binary function (multiplication) and two unary ones.

```
In[20]:= LambdaStar[x][Sin[x] Cos[x]]
Out[20]= S[S[K[c[Times]]][Cos]][Sin]
```

4.4 Converting Functions to Combinators

Now we have all of the tools needed to convert pure functions into combinators. We can simply turn the function Function[x, t] into LambdaStar[x, t]. The result of applying either to an argument a, that is, LambdaStar[x, t][a] or Function[x, t][a], is t /. x -> a. Therefore, we can use the combinator in place of the pure function.

Because the syntax for pure functions allows some variations, we need several rules to cover all cases. We can also deal with functions of several variables by turning Function[{*vars*...}, *body*] into LambdaStar[*vars*...][*body*]. The definition of the conversion operation toCombinators[*expr*] is shown in Listing 4.4–1.

```
toCombinators::usage = "toCombinators[expr] converts all
    pure functions in expr into combinators."
SetAttributes[{leafQ, unevQ}, HoldFirst]
leafQ[expr_] := FreeQ[Hold[expr], HoldPattern[Function[_, _]]]
unevQ[expr_] := leafQ[expr] && FreeQ[Hold[expr], HoldPattern[LambdaStar[__][_]]]

toCombinators[ expr_ ] :=
    expr //. {
        HoldPattern[Function[{}, body_]] :> body,
        HoldPattern[Function[{x__}, body_?unevQ]] :> LambdaStar[x][Unevaluated[body]],
        HoldPattern[Function[x_, body_?unevQ]] :> LambdaStar[x][Unevaluated[body]],
        HoldPattern[Function[{x__}, body_?leafQ]] :> LambdaStar[x][body],
        HoldPattern[Function[x_, body_?leafQ]] :> LambdaStar[x][body]
    }
```

Listing 4.4–1: Part of Combinators.m: Converting functions to combinators

The rules contain a few subtleties, which we want to explain here. The predicates leafQ and unevQ in the patterns guarantee that innermost pure functions are converted first, before the functions containing them. The predicate unevQ[*expr*] is true if *expr* contains no functions or instances of LambdaStar; leafQ[*expr*] is true if *expr* contains no functions. Only the innermost functions contain no other functions; they are the only ones for which these predicates are true the first time the substitution operation //. applies the rules.

We give the predicates the attribute HoldFirst, because we want to apply them to parts of our expressions that are not evaluated, such as bodies of pure functions. A predicate in a pattern, such as Function[x_, body_?*pred*], is tested by first matching the pattern variable body with an expression *expr* and then evaluating *pred*[*expr*]. If *pred* did not have the attribute HoldFirst, *expr* would be evaluated in the normal way as the argument of *pred*.

This example shows the effect just described. Our "predicate" prints out its argument, so we can see exactly what expression it receives as argument. The argument is 2x, rather than x+x.

```
In[21]:= Hold[ x + x ] /. Hold[expr_?Print] :> nothing
2 x
Out[21]= Hold[x + x]
```

To further preserve the argument of the predicate in unevaluated form, it is wrapped in Hold when it is used as the argument of FreeQ. Observe also that we wrap all patterns involving Function or LambdaStar in HoldPattern to prevent their evaluation.

If we omit HoldPattern around Function[_, _] we get an error message, because the pattern Function[_, _] is evaluated (as the argument of FreeQ). Treated as a pure function, it is syntactically wrong, as the error message correctly points out.

```
In[22]:= FreeQ[ Hold[Function[x, x+x]],
                Function[_, _] ]
Function::flpar:
   Parameter specification _ in Function[_, _]
     should be a symbol or a list of symbols.
Out[22]= False
```

There is no need to evaluate this pattern, so it is best wrapped in HoldPattern.

```
In[23]:= FreeQ[ Hold[Function[x, x+x]],
                HoldPattern[Function[_, _]] ]
Out[23]= False
```

The order of conversion of the pure functions is important: If an outer pure function containing another one in its body were converted first, Unevaluated would prevent the code of the inner LambdaStar from ever being called.

During conversion, inner pure functions are turned into (unevaluated) expressions involving LambdaStar. Therefore, if the body of a function contains an instance of LambdaStar, we *must* evaluate it to trigger the code of LambdaStar. (Normally, we do not want to evaluate bodies of functions.) Because the rules turn pure functions into something else, every pure function in our expression will eventually be a leaf; therefore, the conditions do not prevent all functions from being converted.

Let us now see the combinators obtained from some important pure functions. Our rules work nicely to convert the pure functions corresponding to the three basic combinators **S**, **K**, and **I** into their simplest forms.

Here is the identity function.

```
In[24]:= toCombinators[ Function[x, x] ]
Out[24]= I
```

Here is the constant function.

```
In[25]:= toCombinators[ Function[{x, y}, x] ]
Out[25]= K
```

Here is the distributive function.

```
In[26]:= toCombinators[ Function[ {x, y, z}, x[z][y[z]]] ]
Out[26]= S
```

4.4 Converting Functions to Combinators

Some of the resulting combinators have standard names. We shall need them in the next section.

This is a constant function that always returns c.	In[27]:= toCombinators[Function[x, c]] Out[27]= K[c]
Our inside-out conversion order makes sure that scoping rules are observed. This nested function returns its *second* argument, not its first one.	In[28]:= toCombinators[Function[x, Function[x, x]]] Out[28]= K[I]
We can easily check our statement.	In[29]:= %[x][y] Out[29]= y
The **B** combinator represents function composition.	In[30]:= B = toCombinators[Function[{f, g, x}, f[g[x]]]] Out[30]= S[K[S]][K]
Verification is trivial, because we constructed it as a pure function with exactly this body.	In[31]:= B[f][g][x] Out[31]= f[g[x]]
The **C** combinator exchanges the two arguments of a Curried function. The symbol C appears also in the system context; therefore, we force it to be created in the current (global) context in the same way that we did in the package.	In[32]:= `C = S[B[B][S]][K[K]] C::shdw: Symbol C appears in multiple contexts     {Global`, System`}; definitions in context Global` may shadow or be shadowed by other definitions. Out[32]= S[S[K[S[K[S]][K]]][S]][K[K]]
	In[33]:= C[f][x][y] Out[33]= f[y][x]
The **W** combinator doubles its second argument.	In[34]:= W = LambdaStar[x, y][x[y][y]] Out[34]= S[S][K[I]]
	In[35]:= W[f][x] Out[35]= f[x][x]

We started this chapter with a discussion of the problems regarding the use of variables in pure functions. The equivalent formulation with combinators uses no variables; therefore, these problem do not occur.

Here is the Curried form of addition add with which we started this chapter, expressed as a combinator. No variables are left; therefore, no renaming problems can occur.	In[36]:= toCombinators[Function[y, Function[x, x+y]]] Out[36]= S[K[S[c[Plus]]]][K]
This combinatory expression corresponds to add[x] in line In[3] (page 75). It is a function that adds x to its argument.	In[37]:= %[x] Out[37]= S[c[Plus]][K[x]]

Indeed, that is what it does.

```
In[38]:= %[ a ]
Out[38]= a + x
```

The use of `Unevaluated` prevents the evaluation of the body of the pure functions during the conversion.

```
In[39]:= toCombinators[ Function[x, x+x] ]
Out[39]= S[c[Plus]][I]
```

If we had not used `Unevaluated`, the body `x+x` would have turned into `2x` and we would have gotten this combinator. (In this simple case, the two resulting functions behave the same way, however.)

```
In[40]:= toCombinators[ Function[x, 2x] ]
Out[40]= S[K[c[Times][2]]][I]
```

4.5 Applications

Combinatory algebras are mainly of theoretical interest. The first application shows how we can define fixed-point combinators; then, we show how we can embed natural numbers and computable functions in combinatory algebras.

4.5.1 Fixed Points

Our chapter on higher-order functions (Chapter 3) discussed the fixed points of functions. We constructed an operator **Y** such that $\mathbf{Y}(f) = f(\mathbf{Y}(f))$ The same construction works also in combinatory algebras.

We can define **Y** in terms of the combinators introduced in the previous section. **Y** has the property that $\mathbf{Y}f = f(\mathbf{Y}f)$.

```
In[41]:= Y = W[S][B[W][B]];
```

Fixed points are infinite terms. Here we can see how they are built up. The fixed point is an infinite sequence of nested `f`s: `f[f[f[...[]...]]]`. Note that we lowered the recursion limit for this computation using `Block`.

```
In[42]:= Block[{$RecursionLimit=20}, Y[f] ]
$RecursionLimit::reclim: Recursion depth of 20 exceeded.
$RecursionLimit::reclim: Recursion depth of 20 exceeded.
Out[42]= f[f[f[f[f[f[f[f[f[f[f[f[f[f[f[f[f[Hold[S[K[f]][
              S[S[K[f]]][I]]][Hold[I[S[S[K[f]]][I]]]]
              ]]]]]]]]]]]]]]]]
```

Although an attempt to reduce $\mathbf{Y}f$ leads to an infinite computation, we can still prove that $\mathbf{Y}f = f(\mathbf{Y}f)$ by transforming the term according to the leftmost combinators **B**, **W**, and **S**

4.5 Applications

appearing in the reduction of **Y** = **WS**(**BWB**), as follows:

$$
\begin{aligned}
\mathbf{Y}f &= \mathbf{WS}(\mathbf{BWB})f \\
&= \mathbf{S}(\mathbf{BWB})(\mathbf{BWB})f \\
&= \mathbf{BWB}f(\mathbf{BWB}f) \qquad\qquad (4.5\text{--}1) \\
&= \mathbf{W}(\mathbf{B}f)(\mathbf{BWB}f) \\
&= \mathbf{B}f(\mathbf{BWB}f)(\mathbf{BWB}f) \\
&= f(\mathbf{BWB}f(\mathbf{BWB}f)) \qquad\qquad (4.5\text{--}2) \\
&= f(\mathbf{Y}f)
\end{aligned}
$$

where the last step follows from noting that the second term in Line 4.5–2 is equal to Line 4.5–1 and, therefore, to $\mathbf{Y}f$.

The existence of fixed points for all terms means that a negation combinator would also have a fixed point. But a logic in which a term is equal to its negation is inconsistent. It follows that this system cannot form the basis for ordinary logic.

4.5.2 Computations in Combinatory Algebras

Let us show that we can perform computations with integers in a combinatory algebra. We show how to realize the natural numbers in this formalism. It can be shown that all computable functions can be implemented as combinators.

The truth values `true` and `false` can be given as **T** = **K** and **F** = **KI**, respectively. With this definition, the conditional `if` B `then` M `else` N is simply BMN.

All definitions from this subsection are in the file Numerals.m.	`In[43]:= << MathProg`Numerals``
If B is true, we get M.	`In[44]:= T[M][N]` `Out[44]= M`
If B is false, we get N.	`In[45]:= F[M][N]` `Out[45]= N`

Next, we need a pairing construct. With $\mathbf{P} = \lambda^* mnz.zmn$, the term $\mathbf{P}MN$ is an encoding of the ordered pair $\langle M, N \rangle$. The components can be extracted by applying it to **T** and **F**, respectively.

This term represents the pair $\langle M, N \rangle$.	`In[46]:= pair = P[M][N]` `Out[46]= S[S[I][K[M]]][K[N]]`
The two components are extracted easily.	`In[47]:= {pair[T], pair[F]}` `Out[47]= {M, N}`

Now, we can represent numbers. The integer 0 is represented by $\overline{0} = \mathbf{I}$, and an integer $n > 0$ is represented by $\overline{n} = \mathbf{PF}\overline{n-1}$. The function num[n] performs this conversion.

```
num[0] := I
num[n_Integer?Positive] := P[F][num[n-1]]
```

Here is the term that represents the number 5.

In[48]:= **num[5]**

Out[48]= S[S[I][K[K[I]]]][K[S[S[I][K[K[I]]]][K[S[
    S[I][K[K[I]]]][K[S[S[I][K[K[I]]]][K[S[S[I][K[K[I]]]][
    K[I]]]]]]]]]]

These definitions are useful only if we can actually compute with these numbers. Everything has been set up in such a way that the successor function succ and the predecessor function pred are easy to implement. We need also a zero test zero[n] that returns true, if $n = 0$, and false otherwise. Here they are:

```
succ = LambdaStar[x][P[F][x]]
pred = LambdaStar[x][x[F]]
zero = LambdaStar[x][x[T]]
```

Here is the number $\overline{2}$, defined as the twofold successor of $\overline{0}$.

In[49]:= Nest[succ, num[0], 2]

Out[49]= S[S[I][K[K[I]]]][K[S[S[I][K[K[I]]]][K[I]]]]

Applying the predecessor to it twice gets us back to $\overline{0} = \mathbf{I}$.

In[50]:= Nest[pred, %, 2]

Out[50]= I

How can we define functions on numbers, such as addition? The standard recursive definition of *plus* is

$$plus = \lambda xy. \text{ if } x = 0 \text{ then } y \text{ else } plus(x-1)(y+1). \tag{4.5-3}$$

We convert this definition into a combinatory term p whose smallest fixed point is the desired function *plus*:

$$p = \lambda^* fxy. \text{zero}\, xy(f(\text{pred}\, x)(\text{succ}\, y)). \tag{4.5-4}$$

Now, we can set *plus* = $\mathbf{Y}p$. Because $\mathbf{Y}$ is involved, this term does not have a normal form; evaluation would not terminate. (There are other ways to represent numbers as combinatory terms and to convert recursive definitions into combinators that do not have this problem.)

The CD-ROM contains the packages Combinators.m and Numerals.m, as well as the notebook Combinators.nb, with the examples from this chapter.

Chapter Five

Turing Machines

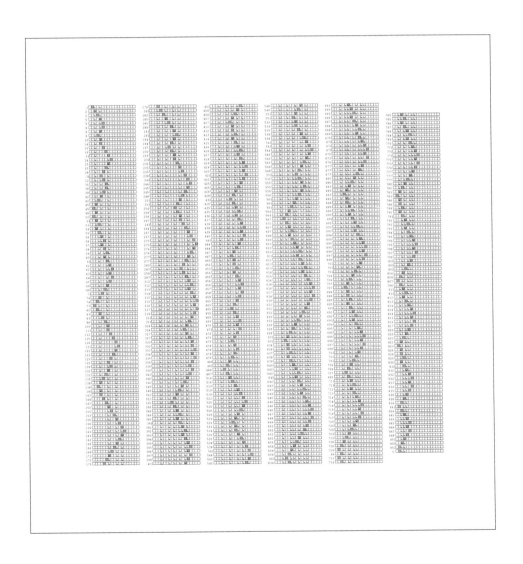

The Turing machine is a simple, yet universal, computing device. Probably every computer science student has at one point written a Turing machine simulator. We develop such a simulator for *Mathematica* in Section 2 and present some tools that make it easier to write programs for Turing machines. In Section 3 we construct an assembler and use it to show explicitly how the primitive recursive functions can be programmed on a Turing machine (Section 4). Finally, in Section 5, we present some code-optimization techniques that are used also in today's RISC machines.

About the illustration overleaf:

The addition 1 + 1 on a Turing machine. The program was developed using the formulation of addition as a primitive recursive function and was then optimized (see Section 5.4). The computation takes 529 steps.

The code for this picture is in BookPictures.m.

5.1 Introduction

The Turing machine (TM) is a model of a computer proposed by Alan Turing in 1936. It has two main properties:

- It is *simple enough* that theorems about the kind of functions computable with it can be proved.
- It is *powerful enough* that any other known computing device can be simulated on it.

The TM consists of a sequence of memory cells, called the *tape*, which is linearly ordered and infinite in both directions. (The theory of computation is not concerned with practical limitations such as memory size and computing time limits, so the TM can have a potentially infinite memory.) There is a finite alphabet of *symbols* that can be written to the individual cells. Only a finite part of the tape will have any values other than "blank" (_) written on it at any time. One memory cell is the *active site*, considered to be the current location of the read/write head. Only this cell can be read and written. The program of the TM is organized much like a finite state automaton. The machine is in one of finitely many *states*, identified by an integer.

One computation step consists of reading the symbol under the head and then performing three actions, depending on that symbol:

- Write a symbol on the tape, overwriting the one already there. The symbol written may be the same as the one read, though.
- Move the head one cell to the left or right, or stay in the same place. We denote this action by l, r, or s.
- Change to a new state.

The cycle then repeats, with the machine reading the symbol under the new location.

The TM is completely described by the instruction set. An instruction set is a table, indexed by states and symbols. Each entry in the table specifies the new symbol to be written, the move, and the new state. The instruction set can be represented as a list of quintuplets of the form

$$\texttt{instruction}[state, symbol, newsym, move, newstate]$$

By convention, the machine halts if it is ever put into a state labeled 0. Also, we shall assume that the machine is started in state 1. If the machine halts because there is no instruction corresponding to the current state and symbol read, we consider this an error condition. We will not (knowingly) prepare instruction sets that can lead to an error condition if the tape contains legal input for the machine.

With this limitation the simplest possible machine has these instructions:

state	symbol	newsym	move	newstate
1	_	*	r	0
1	*	*	r	0

This machine has an alphabet of two symbols, _ and *. It writes * on the tape (independent of what was there before), moves to the right and then stops.

Here is a more interesting TM:

state	symbol	newsym	move	newstate
1	_	_	r	2
2	*	*	r	2
2	_	*	l	3
3	*	*	l	3
3	_	_	s	0

It assumes the head to be initially on an empty cell followed by any number of consecutive marked cells. Can you find out what it does? Let us have a look:

```
1 [ |*|*|*| | | ]
2 [ |*|*|*| | | ]
2 [ |*|*|*| | | ]
2 [ |*|*|*| | | ]
2 [ |*|*|*| | | ]
3 [ |*|*|*|*| | ]
3 [ |*|*|*|*| | ]
3 [ |*|*|*|*| | ]
3 [ |*|*|*|*| | ]
0 [ |*|*|*|*| | ]
```

Each line depicts the configuration before an instruction is executed. On the left we write the current state followed by a portion of the tape. The position of the head is indicated by shading. Thus, the machine starts in state 1 and finds a blank under the head. Accordingly, it leaves the blank alone and moves to the right, changing to state 2. It remains in state 2 moving right while the tape contains marks. At the end it writes another mark and then moves left to where it started. The net effect is to increase the number of consecutive marks by one. We can now see how we can use TMs to do computations on numbers. The integer n can be encoded by n consecutive marks. Our machine implements the *successor function* $n \mapsto n + 1$. We shall encounter this function again when we talk about Turing-computable functions.

The theory of computation uses TMs in two ways. The first use is as *acceptors* of languages. A *language* is a set of words over some alphabet. We write a word on the tape and then start the machine. If it eventually halts in state 0, the word is accepted; if the machine never halts, or halts in an error state, the word is rejected. A language is said to be *Turing recognizable* if a machine can be given that accepts exactly the words belonging to the language. These languages are the type 0 (or recursively enumerable) languages in the Chomsky hierarchy (see [15]).

The second use is to compute functions of integers. We can use a machine to compute a k-place function $f(n_1, n_2, \ldots, n_k)$ by writing the k arguments on the tape, separated by one blank. (The arguments are encoded as sequences of marks, as we have just seen.) If the machine halts in state 0, leaving a consecutive string of n marks on the otherwise empty tape, then it computed

$f(n_1, n_2, \ldots, n_k) = n$. If it halts in any other state, with junk left on the tape, or does not halt at all, the function f is *undefined* for arguments $(n_1, n_2, \ldots, n_k)$. Turing showed that the Turing-computable functions are exactly the partial recursive functions. We shall show part of the proof later.

Before continuing the discussion of Turing computability, let us first write a TM simulator in *Mathematica*.

5.2 A Turing Machine Simulator

To construct a TM simulator we need some auxiliary data structures. First we have to think about how to represent the tape. Because it is unbounded, we cannot use a simple list. Of course, one could argue that we simply make the list "big enough" for all applications. The world is still full of compilers and operating system components with statically sized internal data structures. Eventually some application will reach these bounds. When we ported *Mathematica* to many machines we found such things in some of the lesser C compilers we had to use. By now it should be known to every commercial software developer that such hard limits are not acceptable. Let us therefore search for a truly unbounded data structure for the tape.

5.2.1 Tapes

Our first approach is a straightforward translation of the mathematical model of the tape. The tape is a function from the set of integers into the alphabet of the TM. The function value at i is the symbol in the cell numbered i.

The empty tape is the constant function that always returns _.	`In[1]:= tape[_] = "_"` `Out[1]= _`
To write a symbol s at position i we simply define the function value at this place, for example, at $i = 1$.	`In[2]:= tape[1] = "*"` `Out[2]= *`

With this setup we can use `tape[i]` to look at the value of cell number i. When we overwrite a previously written value, *Mathematica* automatically overwrites the old definition. Internally it uses *hashing* to speed up the lookup of such rules.

This simple approach would work well for our simulation, but it has one drawback: The definitions are *destructive*. We want to keep the execution history of the TM (we already used this idea to produce the picture in Section 5.1). Therefore, we need a nondestructive data structure, which we can update to produce a new one without destroying the old one. *Pure functions* show us a way to do this.

The empty tape is the function that always returns _.

```
In[3]:= emptyTape = Function[i, "_"]
Out[3]= Function[i, _]
```

If we have a tape t and we want to write the value s at position i we get a new tape t' that is identical to the old one except at position i. The new function t' can be described by this formula:

$$t'(j) = \begin{cases} s, & \text{if } j = i \\ t(j) & \text{otherwise} . \end{cases} \tag{5.2-1}$$

The operation update[t, i, s] constructs t' from t.

The result of update[t, i, s] is again a pure function.

```
In[4]:= update[t_, i_, s_] :=
          Function[ j, If[i===j, s, t[j]] ]
```

Here, then, is the abstract data type for the tape:

Constants	emptyTape[b]	empty tape with values b
Constructors	update[*tape*, i, s]	new tape t' with $t'[i] == s$
Selectors	$t[i]$	the value at position i

Note that we can specify the value for blank with an argument of emptyTape.

The implementation of the data type we have just given is not efficient for our particular application. It turns out that we will overwrite old values on the tape much more frequently than we will define new values for previously blank portions of the tape. Therefore, we choose a different implementation of the same abstract data type. It uses a list to store the values. This list is extended as needed. The internal representation is tape[*list*, *offset*, *blank*]. The offset stores the true index of the first element of the list. The code is in **Tape.m**, shown in Listing 5.2–1. Let us see how we can work with such tapes.

We assign an empty tape to the variable t0.

```
In[1]:= t0 = emptyTape["_"]
Out[1]= < _ _ _ _ _ _ ...>
```

Tape t1 contains a mark at position 2. Note the output form we defined for tapes. It prints a portion of the tape in angle brackets.

```
In[2]:= t1 = update[ t0, 2, "*" ]
Out[2]= < _ * _ _ _ _ ...>
```

These tapes behave just like ordinary functions.

```
In[3]:= t1[2]
Out[3]= *
```

5.2 A Turing Machine Simulator

```
(* list[[j]] is tape position i = offset + j *)
emptyTape[b_] := tape[{}, 0, b]
tape[list_, offset_, b_][i_Integer]/; 0 < i-offset <= Length[list] := list[[i-offset]]
tape[list_, offset_, b_][i_Integer] := b
update[ tape[list_, offset_, b_], i_Integer, new_ ] :=
    Which[
        i-offset <= 0,
            tape[ Join[ {new}, Table[b, {offset-i}], list ], i-1, b ],
        i-offset > Length[list],
            tape[ Join[ list, Table[b, {i-offset-Length[list]-1}], {new} ],
                offset, b ],
        True,
            tape[ ReplacePart[list, new, i-offset], offset, b ]
    ]
makeTape[l_List, b_, p0_:1] := tape[ l, p0-1, b]
Format[t_tape] := SequenceForm[ "< ", Infix[Array[t, 6, 1], " "], " ...>" ]
```

Listing 5.2–1: Tape.m: The Turing machine tape

5.2.2 Configurations and Transitions

The *configuration* of a TM is its current state, tape, and position of the head. We keep it in this data structure:

$$\text{config}[\textit{state}, \textit{tape}, \textit{head}]$$

In addition to the obvious selectors we use a selector symbol[*config*], an abbreviation for tape[*config*][head[*config*]] that gives the current symbol under the head.

We have already seen how to specify instructions in the form

$$\text{instruction}[\textit{state}, \textit{symbol}, \textit{newsym}, \textit{move}, \textit{newstate}]$$

Internally we keep them as replacement rules of the form

$$\{\textit{state}, \textit{symbol}\} \rightarrow \{\textit{newstate}, \textit{newsym}, \textit{move}\} \, .$$

This representation allows us to use the built-in pattern matcher to select the instruction that is used in a configuration. Our data type for instructions has selectors named state[], symbol[], newsymbol[], move[], and newstate[] to extract all the components from an instruction. We shall keep all the instructions of a TM in a list.

Executing an instruction transforms the current configuration into a new one, representing the effect of the instruction. The pair (*current state*, *symbol under head*) determines which instruction is executed. Because we represent instructions as rules, we can let replacement make the selection. The right side of the rule is the triplet (*newstate*, *newsym*, *move*). Its values are used to update the configuration. Listing 5.2–2 shows the code of the NextConfiguration function that executes one instruction.

```
NextConfiguration[c_config/; state[c]==0, instructions_] := c
NextConfiguration[c_config, instr_List] :=
    Module[{newstate, newsym, move},
        {newstate, newsym, move} = {state[c], symbol[c]} /. instr;
        If[ Head[newstate] === Symbol,
                Message[ NextConfiguration::noinstr, state[c], symbol[c]];
                Return[c] ];
        config[ newstate,
                update[tape[c], head[c], newsym],
                head[c] + move /. moverules
            ]
    ]
moverules = Dispatch[{r -> 1, l -> -1, s -> 0}]
initialConfiguration[l_List:{}] := config[1, makeTape[l, b], 1]
```

Listing 5.2–2: Part of Turing.m: Interpreting one instruction

The error message is issued if there is no instruction for the current configuration; this is an error condition, as we said earlier. The first rule says that the configuration does not change if the current state is the halt state 0. The move rules translate the symbolic moves r, l, and s into the appropriate displacements of the head. Applying Dispatch[] to these rules speeds up their use in replacements.

The function initialConfiguration[*list*] creates an initial configuration with the tape contents taken from the given list, the first element being tape position one. The machine starts in state one with the head over cell number one.

Running the TM is now a simple matter of setting up an initial configuration and then using either Nest[NextConfiguration[#, *instructions*]&, *config*, *n*] to execute *n* instructions, NestList[] to keep a list of all the intermediate configurations, or FixedPoint[] and FixedPointList[] to let the machine run until it halts.

The function TuringList[*initial*, *instructions*] runs the machine until it halts. An optional third argument can be used to limit the number of steps executed. This is often a good idea, because there are machines that do not halt (and because it is not easy to find out whether one will halt). The initial configuration can either be a config object or a simple list that is passed to initialConfiguration. TuringList returns the list of configurations. The shortcut Turing[] returns the tape after the machine has halted. The definitions are shown in Listing 5.2–3.

The function PlotTuring[*configlist*] plots the given configurations in the way we have already seen. There is a global variable b that contains the value of the default blank symbol _. The variable m contains the default mark * of a two-symbol TM (TMs can have alphabets with more than two symbols, but we shall use only two symbols in all our examples). All these functions are part of the package Turing.m. The picture in Section 5.1 was produced with these commands:

In[1]:= Needs["MathProg`Turing`"]

5.3 Assembly Programming

```
TuringList[c_config, instr_List, n_:Infinity] :=
    Module[{configs},
        configs = FixedPointList[ NextConfiguration[#, instr]&, c, n ];
        If[ Length[configs] <= n + 1,
            Drop[configs, -1],
            configs ]
    ]
TuringList[init_List, instructions_, n_:Infinity] :=
        TuringList[initialConfiguration[init], instructions, n]
Turing[args__] := tape[ Last[TuringList[args]] ]
```

Listing 5.2–3: Part of Turing.m: Running the simulation

Here is the TM program for the successor function.

```
In[2]:= addOne = {
            instruction[ 1, b, b, r, 2],
            instruction[ 2, m, m, r, 2],
            instruction[ 2, b, m, l, 3],
            instruction[ 3, m, m, l, 3],
            instruction[ 3, b, b, s, 0]
        };
```

We started the machine with three marks on the tape.

```
In[3]:= TuringList[ {b, m, m, m}, addOne ] // Short
Out[3]//Short=
    {config[1, < _ * * * _ _ ...>, 1], config[2, <<1>>, 2],
     <<7>>, config[0, < _ * * * * _ ...>, 1]}
```

This command produced the picture on page 90.

```
In[4]:= PlotTuring[ % ];
```

PlotTuring has an option Columns to give the number of columns for rendering the configurations. Because some simulations will be quite long, we often need output in more than one column.

5.3 Assembly Programming

Building instruction sequences for TMs is quite tedious: It closely resembles programming today's computers in machine language. The first tool developed to make this easier is the *assembler*, which automates tasks such as figuring out branch addresses and which allows the definition of *macros*. A macro is an instruction list template that can be expanded to produce instructions for a frequently used auxiliary problem. It is not a subroutine call mechanism, which needs special instructions for saving the return address. The TM does not have such instructions.

A prerequisite for efficient use of macros (and for compilation of high-level languages) is a *calling convention*. This is a specification of how the interface between the calling code and the subroutine or macro should look. Because we shall deal exclusively with machines that implement functions of integers we can assume that the arguments of our macros are consecutive strings of marks. Arguments are separated by exactly one blank and the machine starts on the cell before the first argument.

The first set of tools consists of simple moves. The sequence `right` moves the head one cell to the right, independent of the symbol under the head:

state	symbol	newsym	move	newstate
1	-	-	r	0
1	*	*	r	0

Similarly, `left` moves one cell to the left. `noop` does nothing:

state	symbol	newsym	move	newstate
1	-	-	s	0
1	*	*	s	0

Such a sequence is often needed as a filler in the more complicated constructions seen later.

Next comes a set of tools to assemble instruction lists into longer ones. The most important of these is *composition*. The machine should first execute one program and then, instead of halting, continue with another program. To make this work, two steps are needed: First, the state numbers of the second program need to be changed so as to be disjoint from the states of the first program. Second, every halt state (zero) in the first program needs to be changed to the first state of the second program. This action is often called *code relocation*. The function

$$\texttt{relocate}[\textit{instr, offset, return}]$$

adds *offset* to each state and new state occurring in the instructions, except that new state 0 becomes *return*. We can implement it with two replacement rules:

```
relocate[instr_List, offset_, return_] :=
    instr /. {
            instruction[st_, sy_, nsy_, mv_, 0] :>
                instruction[st+offset, sy, nsy, mv, return],
            instruction[st_, sy_, nsy_, mv_, nst_] :>
                instruction[st+offset, sy, nsy, mv, nst+offset]
        }
```

Now we can compose two instruction lists. We find the maximum state in the first and relocate the second accordingly. Then we change the halt state of the first one to point to the new first state of the second list.

5.4 Recursive Functions

```
maxState[instr_List] := Max[ state /@ instr ]
compose[instr1_, instr2_] :=
    With[{offset = maxState[instr1]},
        Join[ relocate[instr1, 0, offset+1], relocate[instr2, offset, 0] ]
    ]
```

The function splice[*instrlists*...] implements a version of compose with an arbitrary number of arguments. It is a half-line program:

```
splice[ instrs___ ] := Fold[ compose, {}, {instrs} ]
```

As a simple example, here is a right move followed by a left move, which is another, less efficient, version of noop.

```
In[1]:= splice[ right, left ]
Out[1]= {{1, _} -> {2, _, r}, {1, *} -> {2, *, r},
         {2, _} -> {0, _, 1}, {2, *} -> {0, *, 1}}
```

A looping construct also is useful. While the current symbol is a mark, the given program is run once and then the symbol under the head is checked again, until we find a blank and stop:

```
while[ instr_ ] :=
    Join[ { instruction[1, b, b, s, 0], instruction[1, m, m, s, 2] },
          relocate[ ins, 1, 1 ]
    ]
```

With these tools we can write a program that skips over one argument.

First, we move right to put the head under the first mark of the argument. Then we move right while there is still a mark. At the end we will be over the first blank after the argument.

```
In[2]:= skip1 = splice[ right, while[ right ] ]
Out[2]= {{1, _} -> {2, _, r}, {1, *} -> {2, *, r},
         {2, _} -> {0, _, s}, {2, *} -> {3, *, s},
         {3, _} -> {2, _, r}, {3, *} -> {2, *, r}}
```

Of course, the program should also work if the argument is zero. Then there are no marks at all and we should simply move one cell to the right.

```
In[3]:= TuringList[ {b}, skip1 ]
Out[3]= {config[1, < _ _ _ _ _ _ ...>, 1],
         config[2, < _ _ _ _ _ _ ...>, 2],
         config[0, < _ _ _ _ _ _ ...>, 2]}
```

In the next section we shall need a few more manipulations of argument sequences, such as copying arguments to other places on the tape, and erasing an argument and shifting everything to the left so that the next argument is in the place of the erased one. The complete code of all of these macros is in the package TuringMacros.m, reproduced in Listing 5.7–1 at the end of this chapter.

5.4 Recursive Functions

Let us now turn to computable functions. *Church's thesis* says that the functions computable by any (sufficiently powerful) computing device are exactly the *partial recursive functions*. This has been proved for all computing devices or models proposed so far, be they Turing machines, pushdown automata, cellular automata, tag machines, λ-calculus, combinatory logic, and so on. Many of these results appeared almost simultaneously in the 1930s. We now show that the partial recursive functions are Turing computable.

5.4.1 Primitive Recursive Functions

First, we need to look at primitive recursive functions. The primitive recursive (PR) functions are a subclass of the partial recursive functions. Most arithmetic operations and simple Do loops in programs are PR functions. Primitive recursive functions are built-up from other PR functions and a few basic functions. (This process is closely analogous to the definition of data types from constants and constructors.) There are five construction principles. We shall give *Mathematica* code for these, instead of a purely formal treatment. There is a lot of clumsy notation involved in a strictly formal presentation, which obscures the ideas, rather than helps provide an understanding of them.

1. The constant 0 is a (nullary) PR function.

2. The *successor function*
$$\text{succ[m\_] := m + 1}$$
   is a unary PR function.

3. For all $n \geq 1$ and $k \leq n$, the *projection function*
$$\text{p[}k\text{, }n\text{][m1\_, m2\_, ..., m}n\text{\_] := m}k$$
   is PR. It returns argument k of its n arguments. p[1, 1] is the *identity*.

4. The *composition* of PR functions is PR. Let f be a p-ary, and $g_1, g_2, \ldots, g_p$ be n-ary PR functions. Then the function
$$\text{h[m1\_, m2\_, ..., m}n\text{\_] :=}$$
$$f[g_1[\text{m1, m2, ..., m}n], g_2[\ldots], \ldots, g_p[\text{m1, m2, ..., m}n]]$$
   is also PR.

5. *Primitive recursion.* Let f be an n-ary, g an $(n + 2)$-ary PR function, $n \geq 0$. Then h, with
$$\text{h[0, m1\_, m2\_, ..., m}n\text{\_]} := f[\text{m1, m2, ..., m}n]$$
$$\text{h[k\_, m1\_, m2\_, ..., m}n\text{\_] :=}$$
$$g[\text{k-1, h[k-1, m1, m2, ..., m}n], \text{m1, m2, ..., m}n]$$
   is also PR.

5.4 Recursive Functions

The PR functions are *total*, that is, they are defined for all values of their arguments. This can be proved by an easy induction for construction 5 and it is trivially true for the others.

5.4.2 Partial Recursive Functions

Partial recursive functions can be introduced by one more construction, the μ-scheme:

6. Let g be an $(n + 1)$-ary PR function. Then h, with

```
h[m1_, m2_, ..., mn_] := (
    k = 0;
    While[ g[m1, m2, ..., mn, k] != 0, k++ ];
    k)
```

is *partial recursive*. The function value $h[m_1, m_2, \ldots, m_n]$ is the smallest k, such that $g[m_1, m_2, \ldots, m_n, k]$ is zero.

A partial recursive function need not be defined for all values of its arguments. If no value of k exists for which $g[m_1, m_2, \ldots, m_n, k]$ is zero, the loop will never terminate.

5.4.3 Some Recursive Functions

It is usually easy but tedious to show that certain functions are PR or partial recursive. The addition function plus[x, y] for example, can be shown to be PR (Listing 5.4–1).

```
plus[0, m1_]   := p[1, 1][m1]
plus[k_, m1_]  := g[ k-1, plus[k-1, m1], m1 ]
g[m1_, m2_, m3_] := succ[ p[2, 3][m1, m2, m3] ]
```

Listing 5.4–1: Addition as primitive recursive function

We used the primitive recursion scheme for plus and defined the needed auxiliary function g using composition of basic functions (successor and projection). Usually we do not write the projection functions explicitly; they are only a notational inconvenience. We could simply write

```
plus[0, m1_]   := m1
plus[k_, m1_]  := succ[ plus[k-1, m1] ]
```

We will need the strict formula later, when we write TM programs for PR functions. The integer square root is partial recursive:

```
sqrt[m1_] := (
    k = 0;
    While[ (k*k - m1) + (m1 - k*k) != 0, k++ ];
    k )
```

We are assuming that we have already shown that multiplication and subtraction are PR. Incidentally, subtraction is a bit tricky, because there are no negative numbers. The difference $m - n$ for $n > m$ is defined to be 0. This is why we form both differences and add them to see whether k*k is equal to m1. The square root function is defined only for integers which are perfect squares. It will not terminate for other integers.

5.4.4 Programming the Partial Recursive Functions

We have to give TM programs for each of the six constructions to show that partial recursive functions are TM computable. In cases 4–6 we assume that we have the programs corresponding to the PR functions *f* and *g* and we use these programs to construct a program for the new function h. We can assume that all programs obey the following calling conventions:

- Arguments are written onto the tape with exactly one blank between them.
- The machine starts in state 1 with the head on the cell immediately before the first argument.
- The machine does not use the tape to the left of the initial cell.
- When finished, the head is again over the initial cell with the result on the following cells.

We shall make heavy use of the macros introduced in Section 5.3.

The Basic Functions

It is easy to give the programs for zero and the successor (see Listing 5.4–2).

```
zero := noop
succ := splice[
            skip[1],
            {instruction[1, b, m, r, 0]}, (* write 1 *)
            skip[-1]
        ]
```

Listing 5.4–2: Part of TuringRecursive.m: Zero and the successor

The zero function takes no arguments and must write zero marks on the tape, that is, it does not have to do anything at all. The successor skips to the end of its argument and writes another mark,

5.4 Recursive Functions

then skips back. The macro `skip[n]` generates the instructions for skipping n arguments. We discussed the successor function on page 90.

The number of projection functions is infinite, so we write a macro parameterized by n and k to generate `p[k, n]` when needed (Listing 5.4–3).

```
p[k_, n_] :=
  splice[
    splice @@ Table[eat1[j], {j, n, n-k+2, -1}],
    If[ k < n,
         splice[ skip[1],
                 splice @@ Table[eat1[j], {j, n-k, 1, -1}],
                 skip[-1]
         ],
         (* else *)
         noop (* for efficiency *)
    ]
  ]
```

Listing 5.4–3: Part of TuringRecursive.m: Projections

The program first "eats" $k - 1$ arguments, using the macro `eat1[j]` which erases the first of j arguments (it needs to know how many arguments there are so it can shift them to the left). Then we skip over the kth argument (the result) and erase any remaining ones; for the sake of efficiency we do this only if we do not project onto the last argument. Note that the `If` is not a TM instruction; it is used only for conditional code generation, much like `#ifdef` in the C preprocessor.

Let us look at an example; we want to compute `p[2, 3][1, 2, 1]`. The result, the second of the three arguments, should be two marks on the tape.

All code from this section is in the package TuringRecursive.m.

```
In[1]:= Needs["MathProg`TuringRecursive`"]
```

First, we generate the instruction list.

```
In[2]:= (p23 = p[2, 3]) // Short
Out[2]//Short=
  {{1, _} -> {2, _, r}, {1, *} -> {2, *, r},
   {2, _} -> {26, _, s}, <<94>>, {62, *} -> {61, *, l}}
```

This simple function needs quite a few instructions.

```
In[3]:= Length[ p23 ]
Out[3]= 98
```

We let it run with arguments 1, 2, and 1 encoded on the tape. It leaves the second argument, that is, two marks, on the tape.

```
In[4]:= TuringList[ {b, m, b, m, m, b, m}, p23 ] // Short
Out[4]//Short=
  {config[1, < _ * _ * * _ ...>, 1], config[2, <<1>>, 2],
   <<76>>, config[0, < _ * * _ _ _ ...>, 1]}
```

This is the length of the computation.

```
In[5]:= Length[ % ]
Out[5]= 79
```

Here is a picture of this computation that returns the second of three arguments.

```
In[6]:= PlotTuring[ %%, Columns-> 4,
                    DefaultFont->{"Courier", 7} ];
```

Composition

The macro comp[n, f, {g_1, ..., g_p}] generates the necessary code. It first "calls" the functions $g_1, g_2, \ldots, g_p$ on the given n arguments. This works by copying the arguments to the end of the used tape with the head immediately before the copy. Then the code of g_i is invoked, which by the calling convention does not disturb the tape on its left and leaves the result in place of the copy of the arguments. At the end of this process all results g_1[m1, m2, ..., mn], ..., g_p[m1, m2, ..., mn] are on the tape, next to each other. The original arguments are then erased and the code for f is invoked, leaving the final result on the tape. Figure 5.4–1 shows a few stages in this process.

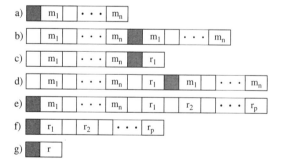

Figure 5.4–1: Composing programs

Line a shows the initial configuration with n arguments on the tape. We copy them to the end, arriving at line b. Now we run the program for g_1 which leaves a result r_1 as shown in line c. Then

5.4 Recursive Functions

we copy the arguments again, ready for running g_2 (line d). After running all of g_1 through g_p the tape looks like line e. We erase the original arguments and are then ready to run f (at line f). The last line shows the final result. The code is given in Listing 5.4–4.

```
comp[ n_, f_, gs_List ] :=
   With[{p = Length[gs]},
     splice[
       splice @@ Table[ splice[ copy[n, n+i-1],  (* copy args *)
                                gs[[i]],         (* call g_i *)
                                skip[-(n+i-1)]   (* back to beginning *)
                              ], {i, 1, p} ],
       splice @@ Table[eat1[i+p], {i, n, 1, -1}],
       f
     ]
   ]
```

Listing 5.4–4: Part of TuringRecursive.m: Composition

The macro copy[n, m] copies the first n out of m arguments to the end of the used tape.

The composition succ[succ[n]] is a function that increments its argument by two.

```
In[7]:= s2 = comp[1, succ, {succ}];
```

We start it with $n = 1$ to get the result 3.

```
In[8]:= TuringList[ {b, m}, s2 ] // Short
Out[8]//Short=
  {config[1, < _ * _ _ _ _ ...>, 1], config[2, <<1>>, 2],
   <<80>>, config[0, < _ * * * _ _ ...>, 1]}
```

Here is a picture of this computation.

```
In[9]:= PlotTuring[ %, Columns-> 4,
                    DefaultFont->{"Courier", 7} ];
```

Recursion

Primitive recursion 5 has to be turned into an iterative process, because we do not have recursion in a TM. We first compute $h_0 = $ h[0, m_1, m_2, ..., m_n] by calling $f[m_1, m_2, ..., m_n]$. In the loop we repeatedly call $h_{i+1} = g[i, h_i, m_1, m_2, ..., m_n]$ with $i = 0, 1, ...$. At each iteration we decrement k until we reach $k = 0$. The details are quite cumbersome. The comments in the code in Listing 5.4–5 show the layout of the tape. The head is marked with : if it is not inside an argument.

```
pr[n_, f_, g_] :=
splice[            (*:k m1...mn                    *)
  skip[n+1],       (* k m1...mn:                   *)
  zero,            (* k m1...mn:0                  *)
  skip[-n],        (* k:m1...mn 0                  *)
  copy[n, n+1],    (* k m1...mn 0:m1...mn          *)
  f,               (* k m1...mn 0:h0               *)
  skip[-(n+2)],    (*:k m1...mn 0 h0               *)
  right,           (* k m1...mn 0 h0               *)
  while[ (* k > 0 *)
    splice[
      left,            (*:k' m1...mn:i hi              *)
      sub1[n+2],       (*:k m1...mn i hi               *)
      skip[n+1],       (* k m1...mn:i hi               *)
      copy[2, 2],      (* k m1...mn i hi:i hi          *)
      skip[-(n+2)],    (* k:m1...mn i hi i hi          *)
      copy[n, n+4],    (* k m1...mn i hi i hi:m1...mn  *)
      skip[-2],        (* k m1...mn i hi:i hi m1...mn  *)
      g,               (* k m1...mn i hi:hi'           *)
      skip[-2],        (* k m1...mn:i hi hi'           *)
      copy[1, 3],      (* k m1...mn i hi hi':i         *)
      succ,            (* k m1...mn i hi hi':i'        *)
      skip[-1],        (* k m1...mn i hi:hi' i'        *)
      copy[1, 2],      (* k m1...mn i hi hi' i':hi'    *)
      skip[-4],        (* k m1...mn:i hi hi' i' hi'    *)
      eat1[5],         (* k m1...mn:hi hi' i' hi'      *)
      eat1[4],         (* k m1...mn:hi' i' hi'         *)
      eat1[3],         (* k m1...mn:i' hi'             *)
      skip[-(n+1)],    (*:k m1...mn i' hi'             *)
      right            (* k m1...mn i' hi'             *)
    ]
  ],
  left,            (*:0 m1...mn k hk                *)
  p[n+3, n+3]      (*:hk                            *)
]
```

Listing 5.4–5: Part of TuringRecursive.m: Primitive recursion

5.4 Recursive Functions

Let us show this program in action with a simple but important program, the *predecessor function*:

$$pred(j) = \begin{cases} 0, & \text{if } j = 0 \\ j - 1 & \text{otherwise} \end{cases} \qquad (5.4\text{--}1)$$

We can recognize it as PR with the following definition:

```
pred[0]  := 0
pred[k_] := p[1, 2][k-1, pred[k-1]]
```

(That is, we throw away the result of the recursive invocation.)

We use the macro for the recursion scheme with f = zero and g = p[1, 2].	In[10]:= pred = pr[0, zero, p[1, 2]];
The sequence is amazingly long.	In[11]:= Length[pred] Out[11]= 722
Here is pred[1] which gives 0. The computation takes 215 steps.	In[12]:= TuringList[{b, m}, pred] // Short Out[12]//Short= {config[1, < _ * _ _ _ _ ...>, 1], config[2, <<1>>, 2], <<212>>, config[0, < _ _ _ _ _ _ ...>, 1]}

For a more complicated example, let us return to addition (see Listing 5.4–1).

The base of the recursion is the identity.	In[13]:= f = p[1, 1];
The recursion step is the composition of the successor and the projection p[2, 3].	In[14]:= g = comp[3, succ, {p[2, 3]}];
The PR scheme is used to define addition.	In[15]:= plus = pr[1, f, g];
Here is the length of the instruction sequence.	In[16]:= Length[plus] Out[16]= 1360
The computation 1 + 1 takes this many steps. The picture on page 87 shows this computation, but with an optimized instruction sequence.	In[17]:= Length[TuringList[args[1, 1], plus]] Out[17]= 835
TuringRecursive.m defines a function args[m_1, ..., m_n] that can be used to create initial tapes with the given arguments.	In[18]:= args[1, 2, 3] Out[18]= {_, *, _, *, *, _, *, *, *}

The μ-Scheme

I leave the construction of the μ-scheme as an exercise.

5.5 Optimization

The long instruction lists and slow execution of the programs just developed beg for optimization. As is often the case, machine programs generated automatically, be it by a compiler or our simple macros, are far from efficient. A number of techniques can be applied to such machine programs to optimize them. We shall apply three such techniques to our TM programs.

5.5.1 Partial Evaluation

The first technique is *partial evaluation*. Looking at the execution histories of the programs for projection and the predecessor, we see many places where the head does not move from one instruction to the next. This comes from the instructions with a move code of s which are introduced by the while macro, the noop instruction, and a few others. If we have an instruction

$$\texttt{instruction}[i, \textit{sym}_1, \textit{sym}_2, \texttt{s}, j],$$

we know that in the next state j we will find the symbol $\textit{sym}_2$ under the tape. If the next instruction is

$$\texttt{instruction}[j, \textit{sym}_2, \textit{sym}_3, \textit{move}, k],$$

the combined effect of these two instructions is

$$\texttt{instruction}[i, \textit{sym}_1, \textit{sym}_3, \textit{move}, k].$$

We can *evaluate* instruction j before the program is actually run. The function merge0 implements this idea; merge applies it to all instructions in a list (see Listing 5.5–1).

We use Cases[*list*, *lhs* :> *rhs*, {1}, 1]] to find all instances of the pattern *lhs* in *list*, transform it into *rhs* and return the list of these. Usually there is just one match. The third and fourth arguments of Cases[] restrict the search to the first level (the elements of *list*), and return at most one matching case. If there is none, there is no next instruction j, so we return the original one. Otherwise, we return the partially evaluated new instruction. Note carefully which arguments of instruction have underscores after them. This is quite subtle.

The optimization techniques from this section are in the package TuringOptimizer.m.

In[19]:= Needs["MathProg`TuringOptimizer`"]

5.5 Optimization

```
merge0[ instr:instruction[i_, sym1_, sym2_, s, j_], ilist_ ] :=
    Module[{next},
        next = Cases[ilist,
                    instruction[j, sym2, sym3_, m_, k_] :>
                        instruction[i, sym1, sym3, m, k],
                    {1}, 1 ];
        If[ Length[next] < 1,
            instr, (* no next instruction *)
            next[[1]]
        ]
    ]
merge0[ i0_, ilist_ ] := i0  (* not mergeable *)
merge[instr_List] := merge0[#, instr]& /@ instr
```

Listing 5.5–1: Part of TuringOptimizer.m: Merging instructions

Let us optimize the predecessor code.	In[20]:= **predfast = merge[pred];**
The computation now takes only 139 steps instead of 215.	In[21]:= **Length[TuringList[{b, m}, predfast]]** Out[21]= 139
The optimization can be repeated, because new redundant instruction pairs may have been formed.	In[22]:= **predfaster = merge[predfast];**
In this example, we gain an additional three cycles.	In[23]:= **Length[TuringList[{b, m}, predfaster]]** Out[23]= 136

5.5.2 Dead Code Removal

The second optimization technique is *dead code removal*. As a consequence of the merging of instructions some instructions may never be used. The reachable instructions can be found like this: Look for all states that can be reached from the first instruction, that is, find all states i such that there is an instruction

$$\text{instruction}[1, \textit{sym}_1, \textit{sym}_2, \textit{move}, i]$$

This is the set S of states reachable after one instruction. Now find all states j such that there is an instruction

$$\text{instruction}[i, \textit{sym}_1, \textit{sym}_2, \textit{move}, j], \quad i \in S$$

and include them in the set S. Continue until S no longer changes. All instructions whose state is not an element of S can now be deleted. This is nothing but the *graph reachability* or *transitive closure* problem. (We can form a directed graph with the states as vertices and edges from vertex i to vertex j if there is an instruction leading from state i to state j.) The algorithm just given is not very efficient. It suffices to compute new states reachable from those states that were added to S

during the last iteration. The code for reach, which computes the reachable states, and for the optimizer deadcode, which removes dead code, is given in Listing 5.5–2.

```
reach[instr_] :=
    Module[{trans, active = {0, 1}, new = {1}, tr0},
        trans = {state[#], newstate[#]}& /@ instr;
        trans = Union[trans]; (* remove duplicates *)
        While[ True,
            tr0 = Select[ trans, MemberQ[new, #[[1]]]& ];
            new = #[[2]]& /@ tr0;
            trans = Complement[trans, tr0];
            new = Complement[new, active];
            If[ new === {}, Break[] ];
            active = Union[ active, new ];
        ];
        active
    ]
deadcode[instr_] :=
    With[{active = reach[instr]},
        Select[ instr, MemberQ[active, state[#]]& ]
    ]
```

Listing 5.5–2: Part of TuringOptimizer.m: Dead code removal

We apply this optimization to the predecessor code.

`In[24]:= predfastest = deadcode[predfaster];`

There were many dead instructions in this list, as can be seen from this comparison of the length of the old and new instruction sequences.

`In[25]:= {Length[predfaster], Length[predfastest]}`
`Out[25]= {722, 383}`

Removal of dead code does not decrease the number of instructions executed, but the simulation will run faster, because the code lists that have to be searched are smaller. In modern computers this means that the *cache memory* is used more efficiently, which also speeds up execution. Such methods are very important for the so-called RISCs (reduced instruction set computers), for example, Sun's SPARC. A reduced instruction set means that more instructions are needed than with a traditional architecture (Motorola 68000 or VAX, for example). But a smaller instruction set can be executed faster, so that, with optimizing compilers, the overall performance is usually better.

We can see that the second sequence executes faster than does the first one.

`In[26]:= {Timing[Turing[{b, m}, predfaster ];],`
`         Timing[Turing[{b, m}, predfastest];]}`
`Out[26]= {{2.42 Second, Null}, {2.03 Second, Null}}`

5.5.3 Hand Tuning

The third optimization technique is *hand tuning*, also called *hacking*. The code for the primitive recursion scheme shows a case where hand tuning would be effective. At the beginning we skip over $n + 1$ arguments to add a zero to the end of the tape and then skip back. Because a zero is literally *nothing*, we could delete the whole sequence from the code. There is nothing wrong with this as long as we know that we will never ever change the representation of numbers in the future. If that happens, and the person who developed the original code is no longer around, we will most likely run into trouble.

We can also hand code certain PR functions completely. In Section 5.4.4, we generated the instructions for addition as a primitive recursive function. The resulting instruction list `plus` had length 1360 (752 optimized). But the following code with 8 instructions also will add two numbers; in fact, it is the shortest one possible:

state	symbol	newsym	move	newstate
1	-	-	r	2
2	*	*	r	2
2	-	*	r	3
3	*	*	r	3
3	-	-	l	4
4	*	-	l	5
5	*	*	l	5
5	-	-	s	0

The idea is to insert a mark in the space between the two arguments and delete a mark at the end.

Here you can see this short program add 2 and 3.

```
In[27]:= PlotTuring[ TuringList[args[2, 3], add],
                    Columns -> 2 ];
```

5.6 Conclusions

The Turing machine is certainly not a viable device for actual computations, but it is close enough to modern computers that similar techniques for assembly programming and code optimization can be developed. The simplicity of the Turing machine makes these techniques easy to understand. The demonstration that partial recursive functions are Turing computable, usually presented rather informally, can be carried out explicitly by constructing the actual programs needed.

We have not mentioned the other direction: showing that every Turing-computable function is partial recursive. To do this we would need to introduce the *universal Turing machine*. Our instruction lists and initial configurations of TMs can be encoded on the tape of another TM. Its program can then simulate the computation performed by the first machine, very much like our own simulator. Through *Gödel numbering*, the whole computation can be expressed as a computation with integers, which can then be shown to be partial recursive. The book [21] contains more about this construction. See also Chapter 6, *Computations with Infinite Structures*, in the first volume of this book for more about recursively enumerable and recursive *sets*. Incidentally we have also shown that *Mathematica* can compute the partial recursive functions.

Turing's work appeared in [57]. On the occasion of the 50th anniversary of this paper there appeared a collection [28] of excellent survey articles on the many fields of theoretical computer science spawned by Turing's and others' work in the 1930s. The place to look for compilation techniques for modern computers is [2].

If you want to play with TMs, a good problem is the *busy beaver*, introduced by T. Rado [50]. A busy beaver of size n is a program with these properties:

- It consists of at most n states, not counting the halt state 0.
- There are just two symbols, as in our examples (blank and mark).
- It is started on an empty tape.
- It eventually halts!
- It leaves as many marks on the tape as possible.

Optimal solutions are known for $n = 1, 2, 3, 4$. For higher values of n no proofs of optimality exist. The best value known so far for $n = 5$ is 4098 marks. Here is the program that achieves this number, found by Heiner Marxen (TU Berlin) in 1989:

state	symbol	newsym	move	newstate
1	_	*	l	2
1	*	*	l	1
2	_	*	r	3
2	*	*	r	2
3	_	*	l	1
3	*	*	r	4
4	_	*	l	1
4	*	*	r	5
5	_	*	r	0
5	*	_	r	3

5.6 Conclusions

You can get an idea of what it does by running it for 200 or so steps (see Figure 5.6–1). It will halt only after 11,798,826 steps.

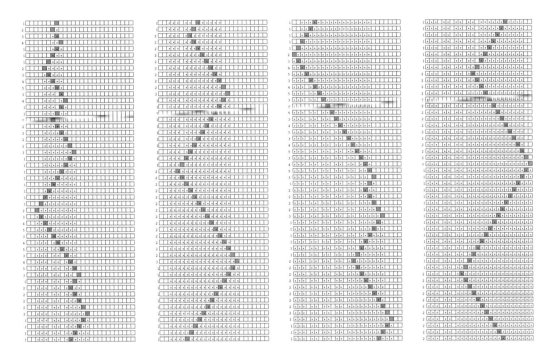

Figure 5.6–1: The alleged busy beaver with five states, first 200 steps

This program came as a big surprise, because the previous maximum for $n = 5$ was 1915 marks. Here is the program that achieved this number, found by G. Uhing in 1986 (see [17, 7]):

state	symbol	newsym	move	newstate
1	_	*	r	2
1	*	*	l	3
2	_	_	l	1
2	*	_	l	4
3	_	*	l	1
3	*	*	r	0
4	_	*	l	2
4	*	*	r	5
5	_	_	r	4
5	*	_	r	2

This program will halt after 2,133,492 steps.

The CD-ROM contains the packages Tape.m, Turing.m, TuringMacros.m, TuringRecursive.m, and TuringOptimizer.m, as well as the notebook TuringMachines.nb, with the examples from this chapter.

5.7 The Complete Code of TuringMacros.m

```
BeginPackage["MathProg`TuringMacros`", "MathProg`Turing`"]

(* assembler for two-symbol Turing machines, {b, m} *)

(* calling convention:
    a macro is an instruction sequence that starts in state 1 and halts in state 0,
    It does not use the tape to the left of its initial position. The square to the
    left of its first argument is empty. Arguments are separated by one empty square.
    The head is initially one square before the first argument.
    It should end in the same position with the result following it.
*)
relocate::usage = "relocate[instr, offset, return] adds offset to all states
    and changes state 0 to return."
splice::usage = "splice[instrlists...] assembles the instruction lists into
    one sequential program."
while::usage = "while[ instrs ] loops over the instructions while the symbol
    under the head is not empty."
(* useful sequences *)
right::usage   = "right moves one step to the right."
left::usage    = "left moves one step to the left."
noop::usage    = "noop performs no instruction."
skip::usage    = "skip[n] skips over n arguments."
copy::usage    = "copy[n, m] copies n arguments over m others and stops before the copy."
shiftleft::usage = "shiftleft[n] shifts n arguments left be one cell."
eat1::usage    = "eat1[n] eats up one argument of n arguments."
(* note: delayed assignment of these lets them pick up changes in b and m *)
Begin["`Private`"]
maxState[{}] = 0
maxState[instr_List] := Max[ state /@ instr ]

relocate[instr_List, offset_, return_] :=
    instr /. { instruction[st_, sy_, nsy_, mv_, 0] :>
                   instruction[st+offset, sy, nsy, mv, return],
               instruction[st_, sy_, nsy_, mv_, nst_] :>
                   instruction[st+offset, sy, nsy, mv, nst+offset] }

splice[ instrs___ ] := Fold[ compose, {}, {instrs} ]

compose[instr1_, {}] := instr1

compose[instr1_, instr2_] :=
    With[{offset = maxState[instr1]},
        Join[ relocate[instr1, 0, offset+1], relocate[instr2, offset, 0] ]
    ]

(* looping *)
```

```
while[ instr_ ] :=
     Join[ { instruction[1, b, b, s, 0],
             instruction[1, m, m, s, 2] },
           relocate[ instr, 1, 1 ]
     ]
(* primitive movements *)
right := {
   instruction[1, b, b, r, 0],
   instruction[1, m, m, r, 0]
}
left := {
   instruction[1, b, b, l, 0],
   instruction[1, m, m, l, 0]
}
noop := {
   instruction[1, b, b, s, 0],
   instruction[1, m, m, s, 0]
}
(* skip arg *)
skip1 := splice[ right, while[ {instruction[1, m, m, r, 0]} ] ]
skipback1 := splice[ left, while[ {instruction[1, m, m, l, 0]} ] ]
(* skip n args *)
skip[n_?Negative] := skipback[-n]
skip[n_] := splice @@ Table[skip1, {n}]
skipback[n_] := splice @@ Table[skipback1, {n}]
(* copy an arg over n others and return after original *)
copy1[n_] :=
splice[ right,
        while[
          splice[
            { instruction[1, m, b, s, 0] }, (* b is sentinel *)
            skip[n+2],
            { instruction[1, b, m, r, 0] }, (* write 1 *)
            skipback[n+2],
            { instruction[1, b, m, r, 0] }] (* restore 1 *)
        ]
]
(* copy n arguments over m and stop before the copy
.arg1.arg2...argn.arg1...argm ->
^
.arg1.arg2...argn.arg1...argm.arg1.arg2...argn.
                                   ^
*)
copy[nn_] := copy[nn, nn]
```

TuringMacros.m

```
copy[nn_, mm_] :=
splice[
    splice @@ Table[copy1[mm-1], {nn}],
    skip[mm - nn]
]
(* shift arg left one pos, ending 2 after the shifted arg *)
(*   .11111.  -->    11111..
     ^                ^
*)
shiftl := [
  instruction[1, b, m, r, 2], (* write 1 *)
  instruction[2, m, m, r, 2], (* skip 1s *)
  instruction[2, b, b, l, 3], (* end *)
  instruction[3, m, b, r, 0]  (* erase 1 *)
}
(* shift n args left by one pos. *)
shiftleft[n_] := splice[ splice @@ Table[ shiftl, {n} ], skipback[n+1] ]
(* eat an argument of n args *)
eat1[n_] :=
splice[ right,
        while[
          splice[
              { instruction[1, m, b, s, 0] }, (* erase it *)
              shiftleft[n],
                right
          ]
        ],
        shiftleft[n-1]
]
End[]
Protect[Evaluate[$Context <> "*"]]
EndPackage[]
```

Listing 5.7–1: TuringMacros.m

Part 2

Visualization

Chapter Six

Animated Algorithms

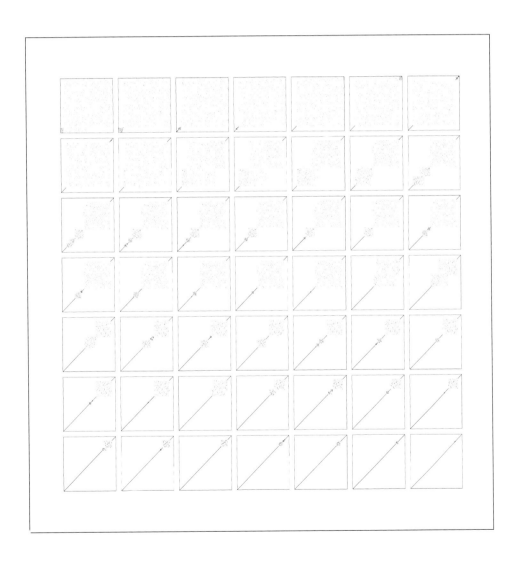

We discuss a method for visualizing the workings of three standard sorting algorithms. The method uses "hooks" in an auxiliary function to obtain run-time data without modifying the implementation of the sorting algorithms themselves.

Section 1 introduces the three algorithms considered: insertion sort, selection sort, and quicksort. Next, we develop the tools for producing graphics and animations that show how the algorithms perform on various inputs. Some more tools are used to investigate the behavior of sorting algorithms for large inputs. We treat them in Section 3. Finally, we mention an improvement to a package for producing static views of animations that we developed for the applications in this chapter.

About the illustration overleaf:
Sorting a random permutation of 1000 elements using quicksort. Shown are 49 snapshots of the list during the sorting process. The tools for producing such graphics are developed in this chapter.

The code for this picture is in BookPictures.m.

6.1 Three Standard Sorting Algorithms

We shall look at three algorithms for sorting arrays: insertion sort, selection sort, and quicksort. A description and analysis of these algorithms can be found in any textbook on algorithms, for example, [53]. All three methods sort an array or list in place, that is, they do not use any auxiliary storage.

Insertion sort and selection sort proceed in a loop over the elements of the array. Both assemble the sorted elements in an initial segment of the array. Insertion sort maintains a sorted initial segment and each subsequent element of the array is inserted into this segment by shifting the elements that are larger to the right. Selection sort repeatedly finds the smallest of the unsorted elements and puts it into its proper place with a single exchange. Quicksort works recursively. In the first phase, the array is partitioned so that all elements in the first part are smaller than all elements in the second part. These two parts can then be sorted independently by two recursive calls of quicksort.

In *Mathematica*, we use lists to hold the elements. The primitive operation is the exchange of two list elements. Parallel assignment allows us to express this exchange in a single statement, without an auxiliary variable. If the variable l holds the list, elements i and j are exchanged by

$$\{l[[i]], \mathtt{l}[[j]]\} = \{l[[j]], \mathtt{l}[[i]]\}.$$

To make our programs more readable, we use an auxiliary procedure $\mathtt{swap}[l, i, j]$ to perform such an exchange. Because we want to modify the value of the parameter l, $\mathtt{swap}$ needs the attribute $\mathtt{HoldFirst}$. (The effect is that its first argument is a reference parameter, see Section 1.1.2.) The code is in the package SortAux.m, reproduced in Listing 6.1–1.

```
BeginPackage["SortAux`"]
swap::usage = "swap[l, i, j] exchanges elements i and j of the value of l."
Begin["`Private`"]
SetAttributes[swap, {HoldFirst}]
swap[l_Symbol, i_, j_] := ({l[[i]], l[[j]]} = {l[[j]], l[[i]]}; l)
End[]
EndPackage[]
```

Listing 6.1–1: SortAux.m: Exchanging elements of a list

We define a list with symbolic elements.	`In[1]:= l = {a, c, b};`
This command exchanges the second and third elements.	`In[2]:= swap[l, 2, 3]` `Out[2]= {a, b, c}`

The code for the three sorting procedures is in the package Sorting.m, shown in Listing 6.1–2. Instead of explaining the code here, I shall develop tools for visualizing the workings of these algorithms in the next section.

```
BeginPackage["Sorting`", "SortAux`"]

insertionSort::usage = "InsertionSort[l] sorts the list l using insertion sort."
selectionSort::usage = "SelectionSort[l] sorts the list l using selection sort."
quickSort::usage = "quickSort[l] sorts the list l using quicksort."

Begin["`Private`"]

insertionSort[list_List] :=
    Module[{l = list, i, n = Length[list], j},
        Do[ j = i-1;
            While[ j >= 1 && l[[j]] > l[[j+1]], swap[l, j, j+1]; j-- ],
            {i, 2, n} ];
        l ]

selectionSort[list_List] :=
    Module[{l = list, i, n = Length[list], min, minj, j},
        Do[ min = l[[i]]; minj = i;
            Do[ If[l[[j]] < min, min = l[[j]]; minj = j], {j, i+1, n} ];
            swap[l, i, minj],
            {i, 1, n-1} ];
        l ]

quickSort[list_] :=
    Module[ {l = list}, qSort[l, 1, Length[l]]; l ]

(* auxiliary procedure *)

SetAttributes[qSort, HoldFirst]

qSort[l_, n0_, n1_] /; n0 >= n1 := l   (* nothing to do *)

qSort[l_, n0_, n1_] :=
    Module[{lm = l[[ Floor[(n0 + n1)/2] ]], i = n0, j = n1},
        While[ True,
            While[ l[[i]] < lm, i++ ];
            While[ l[[j]] > lm, j-- ];
            If[ i >= j, Break[] ];     (* l is partitioned *)
            swap[l, i, j];
            i++; j-- ];
        (* recursion, shorter piece first *)
        If[ i-n0 <= n1-j,
            qSort[ l, n0, i-1 ]; qSort[ l, j+1, n1 ],
            qSort[ l, j+1, n1 ]; qSort[ l, n0, i-1 ] ]
    ]

End[]

EndPackage[]
```

Listing 6.1–2: Sorting.m: Three sorting algorithms

6.2 Sorting in Action

The amount of sorting in a list is easiest to visualize if we restrict ourselves to lists containing a permutation of the integers from 1 to the length of the list. A `ListPlot` of such a list shows the amount of disorder present. A sorted list (the identity permutation) gives a plot with all points along the minor diagonal. Such diagrams appear also in [53].

The auxiliary procedure `PermutationPlot` plots such permutations, choosing good values for the plotting options (see Listing 6.2–1). It is part of the package SortVisual.m, together with the other visualization tools to be developed in this section.

```
PermutationPlot::usage = "PermutationPlot[l, opts...] plots the permutation l."
PermutationPlot[l_List, opts___] :=
    With[{n = Length[l]},
     ListPlot[ l, opts,
         PlotRange -> {{0.5, n+0.5}, {0.5, n+0.5}},
         PlotStyle -> PointSize[0.75/n], Axes -> None,
         FrameTicks -> None, Frame -> True, AspectRatio -> 1 ]
    ]
```

Listing 6.2–1: Part of SortVisual.m: Plotting permutations

Here is the sorted list containing the numbers 1, 2, ..., 20.

`In[1]:= PermutationPlot[ Range[20] ];`

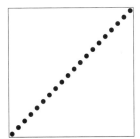

An important test case is a list sorted in reverse.

`In[2]:= PermutationPlot[ Reverse[Range[20]] ];`

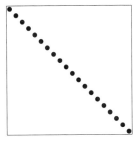

Here is a picture of a random permutation of 20 elements. The function RandomPermutation is defined in the DiscreteMath`Combinatorica` package, taken from [54].

In[3]:= PermutationPlot[RandomPermutation[20]];

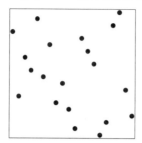

To visualize the progress of a sorting algorithm, we generate a sequence of such permutation plots, one after each exchange that takes place. With most versions of *Mathematica* the resulting sequence of plots can be animated. Here, we'll have to make do with a static version of animation, generated with the package FlipBookAnimation.m.

The remaining problem is to change our implementations of the sorting procedures so that they generate the list of intermediate results. Instead of patching the code of all three procedures, we can modify the auxiliary procedure swap. It serves as our hook into the code. We can change the definition of swap to call another function whenever it is called inside one of the sorting procedures. The function to call is the value of the symbol $swapAction. With suitable values of $swapAction, we can achieve almost any desired effect. To collect data for later analysis or display, we have to use functions with side effects, a programming style normally frowned on.

This technique of providing hooks inside an otherwise unaccessible piece of code is also used by *Mathematica*'s evaluator. You may be familiar with the hooks $Pre, $Post, and $PrePrint. These hooks inside *Mathematica* are described in *The Mathematica Book* [61] in Appendix A.7.

The "hooked" version of swap is given in the file SortAuxG.m (Listing 6.2–2). Note that the context name is still SortAux`. The reason is that the package Sorting.m expects this name in BeginPackage["Sorting`", "SortAux`"]. All we have to do is load our special version first, *before* loading Sorting.m. (Therefore, we restart the kernel at this point.) The default value of $swapAction is the identity function.

As we explained, we must be careful to load the packages in the right order. The package SortVisual.m also contains this sequence of Needs[] commands, so you can simply read it in. Here, we show the commands in detail.

We load the special version by giving an explicit file name.

In[1]:= Needs["MathProg`SortAux`", "MathProg/SortAuxG.m"]

The sorting functions are loaded from Sorting.m.

In[2]:= Needs["MathProg`Sorting`"]

The animation functions are needed, as well.

In[3]:= Needs["Graphics`Animation`"]

6.2 Sorting in Action

```
BeginPackage["SortAux`"]
swap::usage = "swap[l, i, j] exchanges elements i and j of the value of l."
$swapAction::usage = "$swapAction is the function called after each swap."
Begin["`Private`"]
SetAttributes[swap, {HoldFirst}]
swap[l_Symbol, i_, j_] :=
    ({l[[i]], l[[j]]} = {l[[j]], l[[i]]}; $swapAction[l]; l)
End[]
$swapAction = Identity
EndPackage[]
```

Listing 6.2–2: SortAuxG.m: An instrumented version of swapping

This package contains the functions for generating random permutations.

```
In[4]:= Needs["DiscreteMath`Combinatorica`"]
```

Our first application of $swapAction is to collect the intermediate partially sorted lists and to produce permutation plots of the results.

This global variable shall hold all the lists.

```
In[5]:= lists = {};
```

This value of $swapAction appends its argument to lists.

```
In[6]:= $swapAction = AppendTo[lists, #]& ;
```

Here is a random permutation of the numbers 1–9.

```
In[7]:= list = RandomPermutation[9]
Out[7]= {1, 9, 5, 8, 6, 4, 3, 7, 2}
```

We sort it using insertion sort.

```
In[8]:= insertionSort[ list ]
Out[8]= {1, 2, 3, 4, 5, 6, 7, 8, 9}
```

The side effect of $swapAction results in this list of all intermediate permutations.

```
In[9]:= lists // Short
Out[9]//Short=
    {{1, 5, 9, 8, 6, 4, 3, 7, 2}, {1, 5, 8, 9, 6, 4, 3, 7, 2},
     <<19>>, {1, 2, 3, 4, 5, 6, 7, 8, 9}}
```

We convert each of these permutations into a permutation plot (without displaying it).

```
In[10]:= PermutationPlot[#, DisplayFunction -> Identity]& /@
            lists;
```

Here is the animation (at least our static version of it).

In[11]:= ShowAnimation[%];

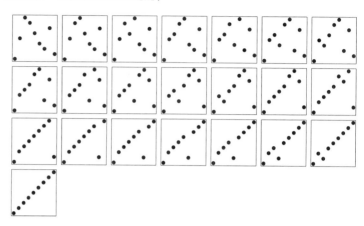

Before we can perform another experiment, we have to reset the global variable.

In[12]:= lists = {};

This time, we use selection sort.

In[13]:= selectionSort[list];

We proceed as before to generate the frames of the animation.

In[14]:= PermutationPlot[#, DisplayFunction -> Identity]& /@
                lists;

In[15]:= ShowAnimation[%];

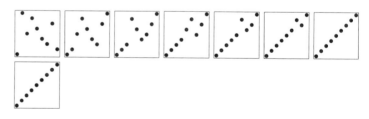

And, finally, for quicksort.

In[16]:= lists = {};

In[17]:= quickSort[list];

In[18]:= ShowAnimation[
            PermutationPlot[#, DisplayFunction -> Identity]& /@
                lists];

At this time, we should think about writing a procedure sortAnimation[*sort*, *list*] that

6.2 Sorting in Action

performs all these steps and produces the animation of *sort[list]* directly. To avoid disturbing the global variable $swapAction, we must use *dynamic binding*, that is, Block[{$swapAction = *value*}, ...]. This is one of the few remaining uses of Block instead of Module. The code declares its own static variable lists inside a Module, then changes the value of $swapAction in the manner indicated and performs the sorting inside of Block. The remaining statements are as before; they produce the list of graphics and animate it (Listing 6.2–3).

```
sortAnimation::usage = "sortAnimation[method, list, (frames), opts...] generates
    an animation of sorting list using method."
sortAnimation[sort_, list_] :=
    Module[{lists = {}},
        Block[{$swapAction = AppendTo[lists, #]&},
            sort[list];
        ];
        ShowAnimation[ PermutationPlot[#, DisplayFunction->Identity]& /@ lists ]
    ]
```

Listing 6.2–3: Part of SortVisual.m: Animating sorting algorithms

We use the new function to investigate the behavior of our three algorithms with a reversed list of length 7. Insertion sort performs poorly.

In[19]:= sortAnimation[insertionSort, Reverse[Range[7]]];

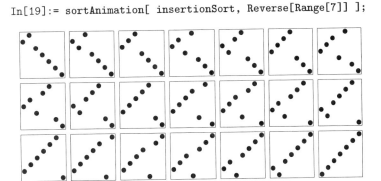

Our version of selection sort performs some trivial exchanges (that is, swap[*l*, *i*, *i*]).

In[20]:= sortAnimation[selectionSort, Reverse[Range[7]]];

Quicksort performs well: It sorts the entire array in the first phase.

In[21]:= sortAnimation[quickSort, Reverse[Range[7]]];

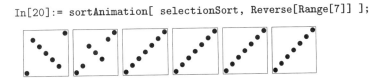

Note that the *Mathematica* frontend allows a much simpler way to generate the animation. You

can simply set $swapAction = PermutationPlot. The graphics will then be created one after another and neatly placed in a grouped sequence of cells. The notebook SortingExamples.nb on the CD-ROM contains examples using this method. The animations are best watched at a very slow speed. On some machines, the horizontal scrollbar can be used to traverse the frames by hand.

After playing around with toy examples, it is time to sort a few "real-world" lists. An animation or collection of pictures for each intermediate step is no longer feasible. Instead, we select a fixed number of equally spaced intermediate values.

The global value of $swapAction is still in effect, so we simply reset the variable lists.

```
In[22]:= lists = {};
```

We generate a random permutation of length 100.

```
In[23]:= list = RandomPermutation[100];
```

Insertion sort is not very efficient, so this computation takes a while.

```
In[24]:= insertionSort[ list ];
```

Here is the number of intermediate steps, that is, the number of exchanges that took place.

```
In[25]:= Length[ lists ]
Out[25]= 2522
```

We select 28 equidistant elements of the list, where the 28 indices are generated with the help of Range.

```
In[26]:= lists[[ Range[ 1, Length[lists],
                Ceiling[Length[lists]/28] ] ]];
```

Here are the 28 frames. The growing sorted initial segment is clearly visible.

```
In[27]:= ShowAnimation[
           PermutationPlot[#, DisplayFunction -> Identity]& /@
             % ];
```

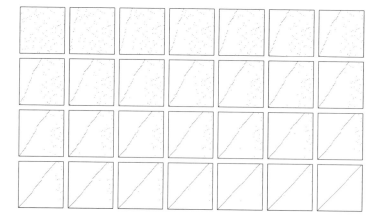

An easy way to program this method is to give a second rule for sortAnimation[*sort, list, k*], where we give the number *k* of frames to display as a third argument. However, the method we

6.2 Sorting in Action

used to select these frames is very space inefficient. First, we generate *all* frames, only to throw most of them away later. The amount of intermediate memory needed to produce an animation for the sorting of a list of n elements grows as fast as $n^2 \log n$ for quicksort and even as fast as n^3 for the other methods.

The solution is not to generate all frames in the first place. If the total number of exchanges is m, and we want to draw k frames, we should keep only every (m/k)th frame. The problem is that we do not know m beforehand. Therefore, we sort the given list twice. The first time we count the number of exchanges, with a method described in Section 6.3; the second time we keep the required frames. The code is given in Listing 6.2–4.

```
sortAnimation[sort_, list_, frames_, opts___?OptionQ] :=
    Module[{lists = {}, swaps = 0, delta},
        Block[{$swapAction = swaps++&},
            sort[list];  (* count number of exchanges *)
        ];
        delta = Floor[swaps/frames];
        swaps = 0;
        Block[{$swapAction = If[Mod[swaps++, delta]==0, AppendTo[lists, #]]&},
            sort[list];  (* collect frames *)
        ];
        If[ Length[lists] > frames, lists = Take[lists, -frames] ];
        ShowAnimation[ PermutationPlot[#, DisplayFunction->Identity, opts]& /@ lists ]
    ]
```

Listing 6.2–4: Part of SortVisual.m: Animating sorting algorithms, version for long lists

Here are 28 frames for quicksort with the same 100-element list. The effect of partitioning can be seen in the blocks forming along the minor diagonal.

In[28]:= sortAnimation[quickSort, list, 28];

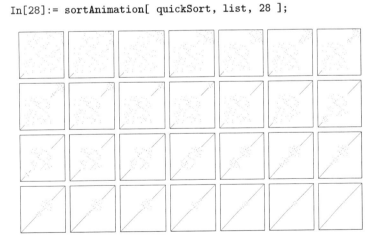

The chapter-opener picture on page 119 shows 49 frames obtained from sorting a 1000-element list with quicksort.

6.3 Asymptotic Behavior

Sorting algorithms have been analyzed thoroughly and are theoretically well understood. Nevertheless, the applications in this section show how one might gather experimental data about less well understood programs.

Sorting a random list of n elements with insertion sort requires on the order of $n^2/4$ comparisons and exchanges. The number of exchanges in selection sort is always equal to $n-1$, but the number of comparisons is again proportional to n^2. Quicksort requires on the order of $n \log n$ comparisons and exchanges. This is the reason that quicksort is the method of choice in most applications.

It is easy to count the number of exchanges using `$swapAction`. We set it to a function that simply counts how often it is called.

```
In[1]:= swaps = 0;\
        $swapAction = swaps++& ;
```

We sort a random permutation of length 200 with quicksort.

```
In[2]:= quickSort[ RandomPermutation[200] ];
```

Here is the number of exchanges needed to sort it.

```
In[3]:= swaps
Out[3]= 345
```

This command performs a number of experiments with random permutations of length n and returns the average number of exchanges performed.

```
Needs["DiscreteMath`Combinatorica`"]
avgSwaps[sort_, n_, trials_] :=
    Module[{swaps = 0},
        Block[{$swapAction = swaps++&},
            Do[ sort[RandomPermutation[n]], {trials} ];
        ];
        N[swaps/trials]
    ]
```

The variable `ntry` gives the number of measurements to be performed, and `nmax` is the highest value of n to use.

```
In[4]:= ntry = 4; nmax = 100;
```

The call of `SeedRandom` guarantees reproducible experiments. You can leave it out if you plan to publish your results in the *Journal of Irreproducible Results* [32].

```
In[5]:= SeedRandom[1];
```

6.3 Asymptotic Behavior

We generate data for lists of length 5, 10, ..., nmax. The result is a list of pairs $\{n, s_n\}$, where s_n is the average number of swaps needed to sort a list of length n.

```
In[6]:= Table[ {n, avgSwaps[insertionSort, n, ntry]},
              {n, 5, nmax, 5} ] // Short
Out[6]//Short=
 {{5, 4.75}, {10, 22.25}, {15, 46.}, {20, 98.25},
  {25, 172.}, <<13>>, {95, 2240.25}, {100, 2560.}}
```

Here is a graphic representation of the data for insertion sort.

```
In[7]:= ListPlot[ %, AxesOrigin -> {0,0},
                PlotRange -> {{0,nmax}, All} ];
```

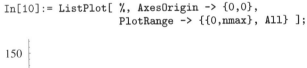

Here are the calculations and results for quicksort.

```
In[8]:= SeedRandom[1];

In[9]:= Table[ {n, avgSwaps[quickSort, n, ntry]},
              {n, 5, nmax, 5} ];
```

Note the different vertical scale in this picture!

```
In[10]:= ListPlot[ %, AxesOrigin -> {0,0},
                 PlotRange -> {{0,nmax}, All} ];
```

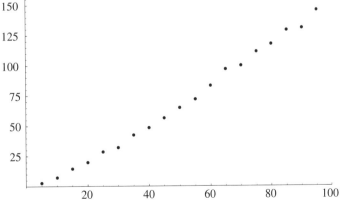

6.4 Conclusions

The three sorting methods from Sorting.m are taken from my textbook [37, Chapter 6]. It contains another method to visualize quicksort that shows the recursive subdivisions, which are more important for its performance than the number of exchanges. To conclude this chapter, here is one more idea for analyzing algorithms, and a discussion of an improvement of the standard package FlipBookAnimation.m

6.4.1 The Built-In Sorting Method

Can we analyze the built-in sorting method, Sort? It does not perform exchanges of the form used earlier; it simply reassigns pointers to some internal data structure. There is one hook: We can give it our own ordering predicate.

We use this variable to count the number of comparisons performed.	`In[11]:= compar = 0;`
Here is an ordering predicate that increments the counter and then behaves like the standard ordering function.	`In[12]:= myOrder[e1_, e2_] := (compar++; OrderedQ[{e1, e2}])`
We sort a random permutation of 1000 elements.	`In[13]:= Sort[ RandomPermutation[1000], myOrder ];`
Here is the number of comparisons performed.	`In[14]:= compar` `Out[14]= 8716`

6.4.2 Static Views of Animations

The static views of the animations in this chapter required an improved version of the FlipBookAnimation.m package, shown in Listing 6.4–1. It is now part of *Mathematica*, Version 3. The original code distributed with Version 2.2 of *Mathematica* could not display lists of graphics with a prime number of frames in a reasonable way. When I developed it, I was not aware of the fact that GraphicsArray (used to assemble the frames in one picture) could cope with an incomplete last row of frames. As a consequence, lists of frames whose length had no proper divisors were displayed in a single row. The new code defines a global variable, Graphics`Animation`$Columns, which can be used to force the number of columns displayed to a certain value. We used it for the images in this chapter.

 The CD-ROM contains the packages Sorting.m, SortVisual.m, SortAux.m, and SortAuxG.m, as well as the notebook SortingExamples.nb, with the examples from this chapter.

6.4 Conclusions

```
Graphics`Animation`$Columns::usage = "Graphics`Animation`$Columns specifies the
        number of columns in the array of animation frames."
Begin["System`"]
$RasterFunction = Identity
$AnimationFunction = Graphics`Animation`MakeGraphicsArray
If[ !ValueQ[Graphics`Animation`$Columns], Graphics`Animation`$Columns = Automatic ]
Begin["Graphics`Animation`Private`"]
Graphics`Animation`MakeGraphicsArray[pics_] :=
    Module[{l = Length[pics], r, row, div, picts},
        If[ l > 1,
            div = Divisors[l];
            r = First[ Select[div, # >= Sqrt[l]&] ];
          , r = 1
        ];
        Which[
            IntegerQ[Graphics`Animation`$Columns] && 1 <= Graphics`Animation`$Columns,
                row = Graphics`Animation`$Columns,
            l <= 2,            (* trivial *)
                row = l,
            r < 1.4 Sqrt[l], (* can divide exactly *)
                row = r,
            True,              (* cannot partition exactly *)
                row = Ceiling[Sqrt[l]];
        ];
        If[ l < row, (* fill in rest with dummies *)
            picts = Join[pics, Table[Graphics[{}], {row-l}]];
          , picts = Partition[pics, row];
            If[ Mod[l, row] > 0,    (* leftovers *)
                AppendTo[picts, Take[pics, -Mod[l, row]]]
            ];
        ];
        Show[GraphicsArray[picts], DisplayFunction -> $DisplayFunction]
    ]
End[] (* Graphics`Animation`Private` *)
End[] (* System` *)
```

Listing 6.4–1: FlipBookAnimation.m: The new code for static animations

Chapter Seven

Function Iteration and Chaos

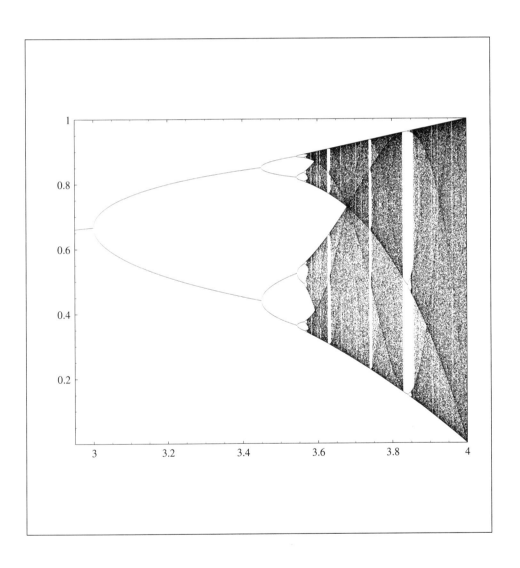

We present a number of programs to investigate and visualize chaos as it occurs with an iterated application of functions. We look at ways to picture orbits under repeated application of a function (in Section 1) and to draw final-state diagrams (in Section 3). We discuss symbolic and numerical methods to find periodic orbits and bifurcation points (Section 2), as well as super-attractive orbits and the Feigenbaum constant (Section 5). Further topics include statistical analysis and visualization of chaotic phenomena such as sensitivity, mixing, ergodic orbits, and intermittency (in Section 4).

About the illustration overleaf:

The picture shows the final-state (or Feigenbaum) diagram for the logistics map $x \mapsto rx(1-x)$ for $2.95 \leq r \leq 4$. Algorithms for producing such pictures are developed in Section 7.3.

The code for this picture is in BookPictures.m.

7.1 Function Iteration

If a real-valued function f maps the unit interval $[0, 1]$ (or any other interval) into itself, the sequence of iterates
$$x_{i+1} = f(x_i), \qquad i = 0, 1, \ldots$$
will always stay in that interval. The sequence of points (x_i) is called the *orbit* of the point x_0. The study of such iteration sequences shows many of the main ingredients of chaos.

Our main example of an iterated function is the *logistics map*, that is, the transformation $f_r : x \mapsto rx(1-x)$, where $1 \leq r \leq 4$. The behavior of its iterates depends on the parameter r and shows large qualitative changes for different ranges of r.

LogisticsMap[r] is a higher-order function that generates the function f_r.

```
In[1]:= LogisticsMap[r_] := Function[x, r x (1-x)]
```

Here is the logistics map f_4. To avoid conflicts of names, *Mathematica* renames the formal parameter of a pure function when substitutions are made in the function's body.

```
In[2]:= f4 = LogisticsMap[4]
Out[2]= Function[x$, 4 x$ (1 - x$)]
```

Orbits can easily be produced with NestList. Here, the initial point is $x_0 = 0.3$ and we perform 15 iterations.

```
In[3]:= NestList[f4, 0.3, 15]
Out[3]= {0.3, 0.84, 0.5376, 0.994345, 0.0224922, 0.0879454,
         0.320844, 0.871612, 0.447617, 0.989024, 0.0434219,
         0.166146, 0.554165, 0.988265, 0.0463905, 0.176954}
```

In the following sections we develop tools to visualize orbits, investigate the phenomena that happen as we increase the value of the parameter r of the logistics map, and then visualize some of the ingredients of chaos.

All commands to be developed in this chapter are in the package IteratedFunctions.m, reproduced in part throughout this chapter.

```
In[1]:= Needs["MathProg`IteratedFunctions`"]
```

7.1.1 Graphical Iteration

A simple technique to visualize function iteration is *graphical iteration*. Start with an initial value x_0. To find the next value $x_1 = f(x_0)$, draw a vertical line from $\{x_0, 0\}$ to $\{x_0, x_1 = f(x_0)\}$. Then, draw a horizontal line to meet the diagonal line $y = x$ at $\{x_1, x_1\}$. Next, draw a vertical to $\{x_1, x_2 = f(x_1)\}$, and repeat the process for each iteration.

Here are the first 20 iterates of $f(x) = 2.9x(1-x)$ with initial value $x_0 = 0.5$.

`In[2]:= FunctionIteration[Function[x, 2.9 x(1-x)], 0.5, 10];`

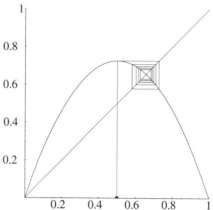

Let us look at the code of `FunctionIteration[`f`, {`x_{min}`, `x_{max}`}, `x_0`, `l`, `$opts$`...]` (excerpted in Listing 7.1–1). Note that the x range $\{x_{min}, x_{max}\}$ has been declared optional; the default value is `{0, 1}`, which occurs quite frequently. The main task is to construct the list of pairs $\{\{x_0, 0\}, \{x_0, x_1\}, \{x_1, x_1\}, \{x_1, x_2\}, \{x_2, x_2\}, \ldots\}$. We start with the list $\{x_0, x_1, x_2, \ldots\}$, easily created with `NestList[`f`, `x_0`, `l`]`. A symbolic example shows the remaining steps.

Here is a typical list of values obtained from `NestList`.

`In[3]:= o = {x0, x1, x2, x3, x4};`

Transposing two copies of o gives us a list in which each entry occurs twice.

`In[4]:= Transpose[{o, o}]`
`Out[4]= {{x0, x0}, {x1, x1}, {x2, x2}, {x3, x3}, {x4, x4}}`

We remove the inner list braces.

`In[5]:= Flatten[ % ]`
`Out[5]= {x0, x0, x1, x1, x2, x2, x3, x3, x4, x4}`

Overlapping partitions of size 2 give us the required result, ...

`In[6]:= Partition[ %, 2, 1 ]`
`Out[6]= {{x0, x0}, {x0, x1}, {x1, x1}, {x1, x2}, {x2, x2},`
`        {x2, x3}, {x3, x3}, {x3, x4}, {x4, x4}}`

... except at the beginning. The first point is special and is modified "by hand."

`In[7]:= ReplacePart[ %, 0, {1, 2} ]`
`Out[7]= {{x0, 0}, {x0, x1}, {x1, x1}, {x1, x2}, {x2, x2},`
`        {x2, x3}, {x3, x3}, {x3, x4}, {x4, x4}}`

This computation is performed in the auxiliary procedure `OrbitToPoints`. The resulting list can be packed into `Line` to generate the graphical iteration.

The remaining code in `FunctionIteration` handles the option `PreIterations`, which specifies the number of iterations to perform invisibly, plots the function f, and draws the bisector $y = x$. The standard trick of setting `DisplayFunction` to `Identity` is used to generate

7.1 Function Iteration

```
FunctionIteration::usage = "FunctionIteration[f, {xmin, xmax}, x0, length, opts..]
    shows the graphics iteration of f[x] with initial point x0.
    length iterations are shown. The default x range is {0, 1}."
Coloring::usage = "Coloring -> f is an option of FunctionIteration that specifies
    the color (or plot style) to use for the orbits. f[i, n] is called to find
    the style of the ith out of n orbits. The default uses different hues."
PlotStyle::usage = PlotStyle::usage <>
    " It gives also the styles for the points and lines of FunctionIteration."
Options[FunctionIteration] = {
    PreIterations -> 0,
    Coloring -> Automatic,
    PlotStyle -> {Thickness[.001], PointSize[0.02]}
};
FunctionIteration[map_, xr:{xmin_, xmax_}:{0, 1}, x0_List, length_, opts___?OptionQ] :=
    Module[{x1, orbits, plot, lines, x, n = Length[x0], ps, pi},
        pi = PreIterations /. {opts} /. Options[FunctionIteration];
        ps = PlotStyle /. {opts} /. Options[FunctionIteration];
        ps = Sequence @@ Flatten[{ps}]; (* must splice in *)
        x1 = Nest[map /@ # &, x0, pi];
        orbits = Transpose[ NestList[map /@ # &, x1, length] ];
        plot = Plot[map[x], {x, xmin, xmax}, DisplayFunction -> Identity];
        col = Coloring /. {opts} /. Options[FunctionIteration];
        If[ col === Automatic, col = coloring ];
        lines = MapIndexed[{col[#2[[1]], n], Line[OrbitToPoints[#1]]}&, orbits];
        Show[ plot,
            Graphics[{ps, lines, Point[{#, 0}]& /@ x1,
                    Line[{{xmin, xmin}, {xmax, xmax}}]}],
            FilterOptions[Graphics, opts],
            AxesOrigin -> {xmin, xmin}, DisplayFunction -> $DisplayFunction,
            PlotRange -> {{xmin, xmax}, {xmin, xmax}}, AspectRatio -> Automatic
        ]
    ]
FunctionIteration[map_, xr:{_, _}:{0, 1}, x0_, args__] :=
    FunctionIteration[map, xr, {x0}, args]
OrbitToPoints[o_List] :=
    ReplacePart[ Partition[Flatten[Transpose[{o, o}], 1], 2, 1], 0, {1, 2} ]
coloring[1, 1] := Hue[0, 0, 0]
coloring[i_, n_] := Hue[ (i-1)/n ]
```

Listing 7.1–1: Part of IteratedFunctions.m: Graphical iteration

the plot of f without displaying it. The computation of the orbit looks a bit more complicated than just explained. The reason is that we allow a *list* of initial points. Different colors are automatically chosen for the iterates of the various initial points. The main reason to use several initial points is to demonstrate the sensitivity property of chaotic systems, which is done in Section 7.4.1.

To assign different colors to the elements of a list of graphics objects, we use a function `coloring[i, n]`, which defines a `Hue` color directive for the ith out of n objects. A special rule for `coloring[1, 1]` makes sure that the image stays monochrome if only one initial point is given. The default coloring function can be overridden with the option `Coloring -> col`, where *col* should be a function of two arguments, `col[i, n]`, that gives the graphic directives for rendering the ith out of n orbits. A suitable value of this option for grayscale images is `Function[{i,n}, GrayLevel[0.9(i-1)/n]]`. The first argument i of the coloring function is equal to the position of the respective graphics object in the list of all objects to be colored. The operation `MapIndexed[g, list]` is used to pass each element of a list, along with its index, to the function g (see Section 1.3.2). A symbolic example should make this technique clear.

Let this be our list of graphics objects.	`In[8]:= list = {g1, g2, g3, g4};`
We find its length.	`In[9]:= n = Length[list]` `Out[9]= 4`
Like `Map`, `MapIndexed[g, list]` applies the function g to each element of the list, but it also passes the index of the element as a second argument to g (note the extra braces).	`In[10]:= MapIndexed[ g, list ]` `Out[10]= {g[g1, {1}], g[g2, {2}], g[g3, {3}], g[g4, {4}]}`

In place of g we use a pure function that extracts the index from its second argument and passes it to our coloring function. The result is packed into a list to restrict the effect of the coloring directive to the following graphics object.	`In[11]:= MapIndexed[ {col[#2[[1]], n], #1}&, list ]`
	`Out[11]= {{col[1, 4], g1}, {col[2, 4], g2},`
	`  {col[3, 4], g3}, {col[4, 4], g4}}`

Examples of the use of color are shown in Plates 1-a and 1-b. Note that the result of the coloring function need not be a color; any graphic directive that sets the style of a line can be used, such as `Dashing` or `Thickness`. (The chapter-opener graphic on page 57 shows the use of different dashing.)

Here is an example of the use of `FunctionIteration` with several initial points given in a list. We want to plot the orbits of a number of nearby initial points.

We generate 20 points in an interval of size 0.01.	`In[12]:= 0.2 + 0.01 Range[20]/20`
	`Out[12]= {0.2005, 0.201, 0.2015, 0.202, 0.2025, 0.203,`
	`   0.2035, 0.204, 0.2045, 0.205, 0.2055, 0.206, 0.2065,`
	`   0.207, 0.2075, 0.208, 0.2085, 0.209, 0.2095, 0.21}`

7.1 Function Iteration

The logistics map $f_4(x) = 4x(1-x)$ is chaotic. Already after five iterations, the values have spread over a large interval. An animation of this dispersion can easily be created; see the CD-ROM supplement for more examples. The coloring function generates gray levels, appropriate for this printed medium.

```
In[13]:= FunctionIteration[ LogisticsMap[4.0], %, 5,
            Coloring ->
               Function[{i,n}, GrayLevel[0.9(i-1)/n]]
         ];
```

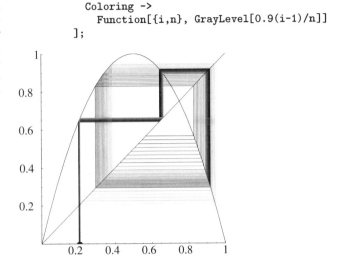

For $r = 3.5$ an attractive four-cycle exists. To better visualize the cycle, we perform a number of invisible iterations to get the initial point close to the cycle before starting the graphical iteration.

```
In[14]:= FunctionIteration[ LogisticsMap[3.5], 0.1, 10,
            PreIterations -> 20 ];
```

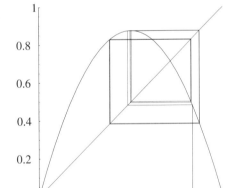

The remaining option `PlotStyle` is used to set the overall styles of the lines and initial points. This option is used by many built-in graphic commands as well.

7.1.2 Time Series

Another often-used visualization of function iteration is the *time series*. It is a plot of x_i versus i. In *Mathematica*, it translates to a simple list plot, see Listing 7.1–2.

```
TimeSeries::usage = "TimeSeries[f, x0, n, opts..] plots the time series
    if n iterations of f[x] starting with x0."

PreIterations::usage = "PreIterations -> n is an option of FunctionIteration and
    TimeSeries that specifies the number of invisible
    iterations that are performed before graphics output is generated."

Options[TimeSeries] = {
    PreIterations -> 0
};

TimeSeries[map_, x0_, length_, opts___?OptionQ] :=
    Module[{pi, x},
        pi = PreIterations /. {opts} /. Options[TimeSeries];
        x = Nest[map, x0, pi];
        ListPlot[ NestList[map, x, length-1],
                FilterOptions[ListPlot, opts], PlotJoined -> True ]
    ]
```

Listing 7.1–2: Part of IteratedFunctions.m: Time series

Here is a time series of the chaotic logistics map showing 100 iterations.

In[15]:= TimeSeries[LogisticsMap[4], 0.1, 100];

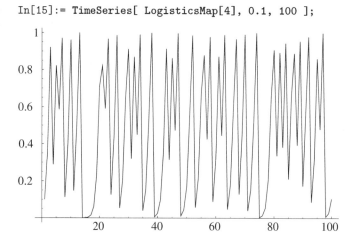

In Section 7.4.5, we use the time series to demonstrate intermittency, one of the phenomena that can appear with chaotic iteration.

7.2 Bifurcations

The logistics map $f_r(x) = rx(1-x)$ has two fixed points in the interval $0 \leq x \leq 1$ for $1 < r \leq 4$. (The fixed points occur where the graph of f_r intersects the line $y = x$.) The fixed point $x = 0$

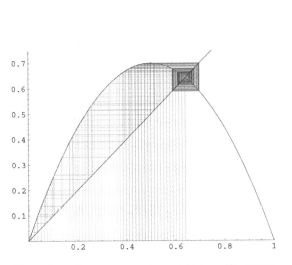

Plate 1-a: Convergence toward the single attractive fixed point of the logistics map $f_{2.8}$. Shown are six iterations of 50 initial points chosen on the left of the fixed point (Section 7.1.1).

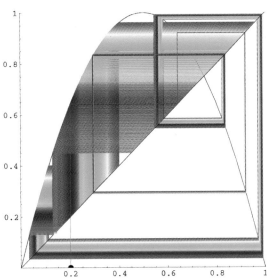

Plate 1-b: The sensitivity of initial conditions of the logistics map f_4. Shown are 500 initial points in an interval of length 0.001 and their first 9 iterations (Section 7.1.1).

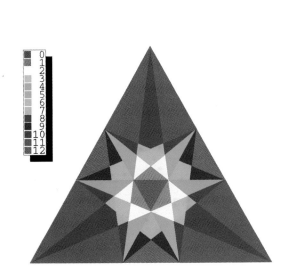

Plate 2-a: The facets from which the stellated icosahedra are constructed (Section 10.2.1).

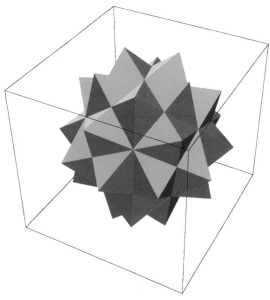

Plate 2-b: A stellated icosahedron with a unique color for each plane (Section 10.3).

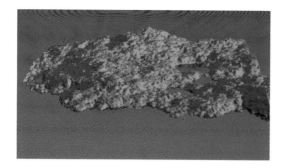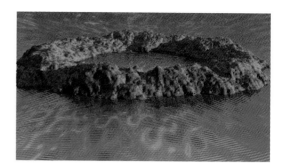

Plate 3-a: Two landscapes generated by fractional Brownian motion and rendered with POVRAY. *Left:* a surface from Section 8.2.1 with dimension 2.25. *Right:* a crater derived from a Bessel function, see Section 8.2.2.

Plate 3-b: A rocky mountainous terrain with a granite texture. The height data were obtained with Fourier synthesis (Section 8.3).

Plate 4-a: The great icosidodecahedron as a stereo pair (Section 11.5).

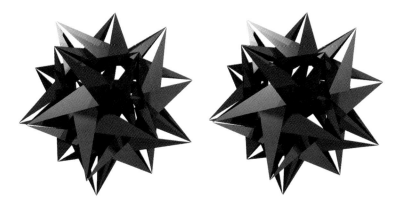

Plate 4-b: A stellated icosahedron as a stereo pair (Section 11.5).

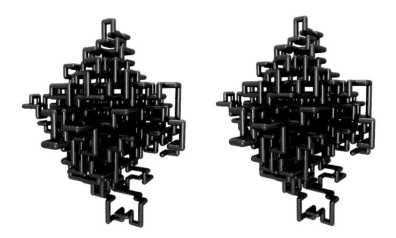

Plate 4-c: A biased, self-avoiding random walk (Section 11.5).

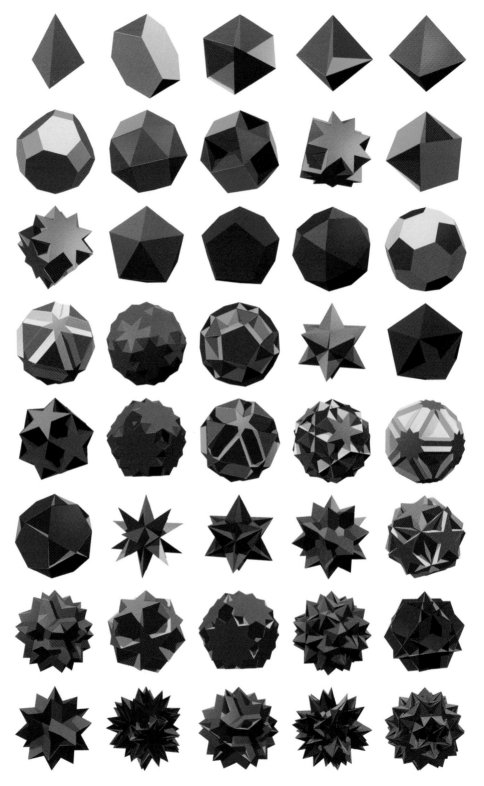

Plate 5: All 75 uniform polyhedra, as well as 5 examples of prisms. The solids are arranged from left to right and from top to bottom across two pages (Section 9.4).

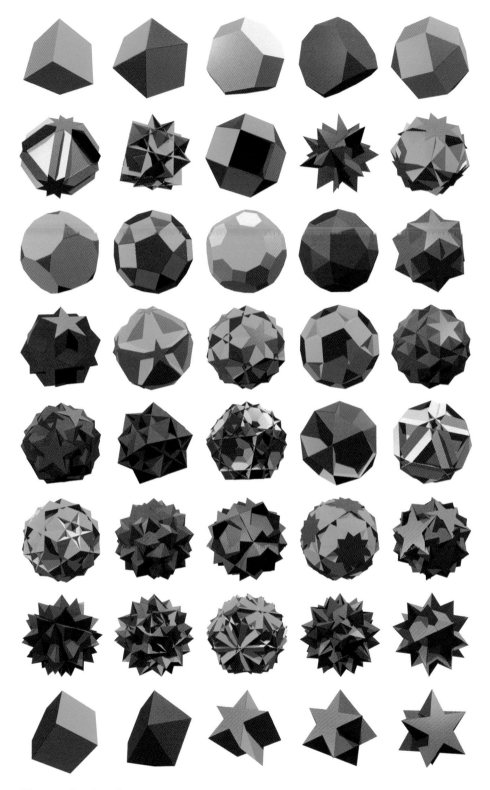

Plate 5: Continued.

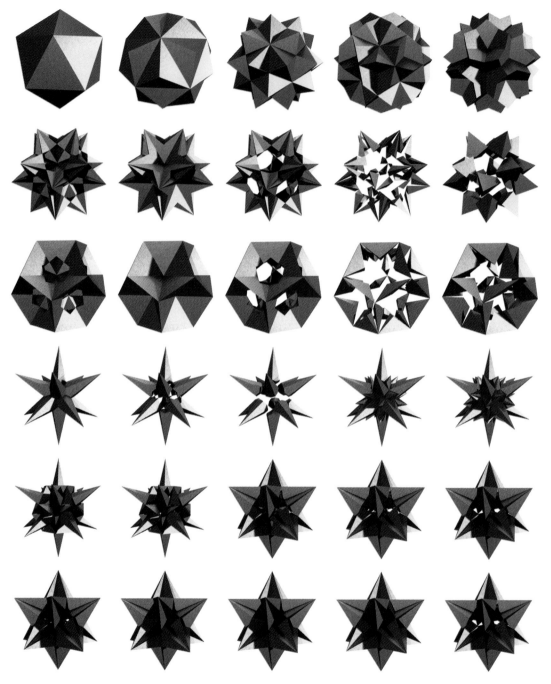

Plate 6: The 59 stellated icosahedra, ordered by increasing size. The solids are arranged from left to right and from top to bottom across two pages (Section 10.3).

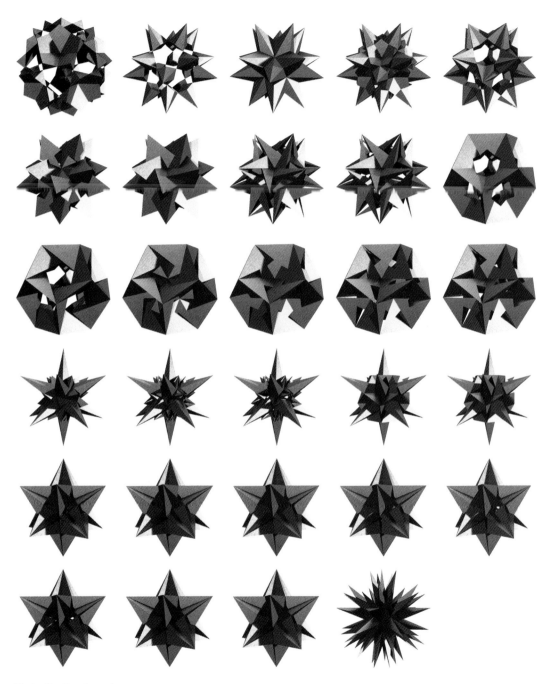

Plate 6: Continued.

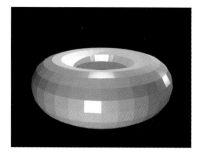

Plate 7-a: A torus rendered in *Mathematica* (Section 11.1).

Plate 7-b: The torus rendered with POVRAY, flat shading (Section 11.2).

Plate 7-c: The torus with smooth shading.

Plate 7-d: The torus as part of a scene.

Plate 8: The great dirhombicosidodecahedron ($|\frac{3}{2}\ \frac{5}{3}\ 3\ \frac{5}{2}$), the only non-Wythoffian uniform polyhedron (Section 9.4.2).

Plate 9: The *Mathematica* icon floating in water (Section 11.4.1).

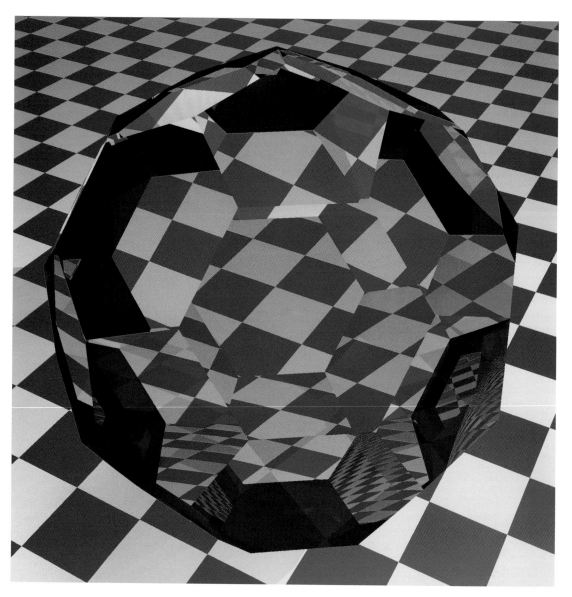

Plate 10: A crystal truncated icosahedron (soccer ball) (Section 11.4.2).

Plate 11-a: A minimal surface (Section 11.4.3).

Plate 11-b: One of C. Henry Edwards' twisted tubes (Section 11.4.4).

Plate 12: A shell (Section 11.4.5).

Plate 13: A wallpaper stereogram. Shown are three uniform polyhedra, from top to bottom: the small dodecicosidodecahedron 3/2 5|5, the great icosidodecahedron 2|5/2 3, and the inverted snub dodecadodecahedron |5/3 2 5. (Image courtesy of F. Bachmann, ETH Zurich, renderings of the solids provided by the author.)

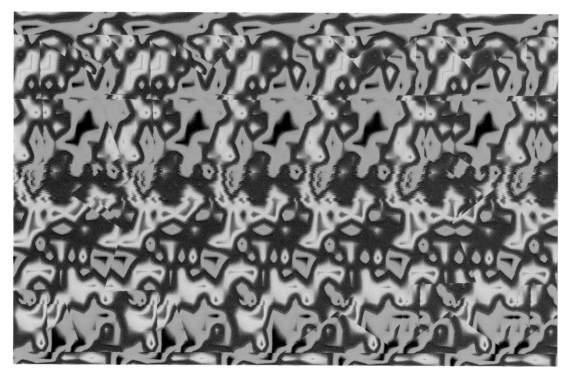

Plate 14-a: The great icosidodecahedron rendered with RAYsis (Section 12.5.1).

Plate 14-b: A twisted tube with a fractional pattern (Section 12.5.2).

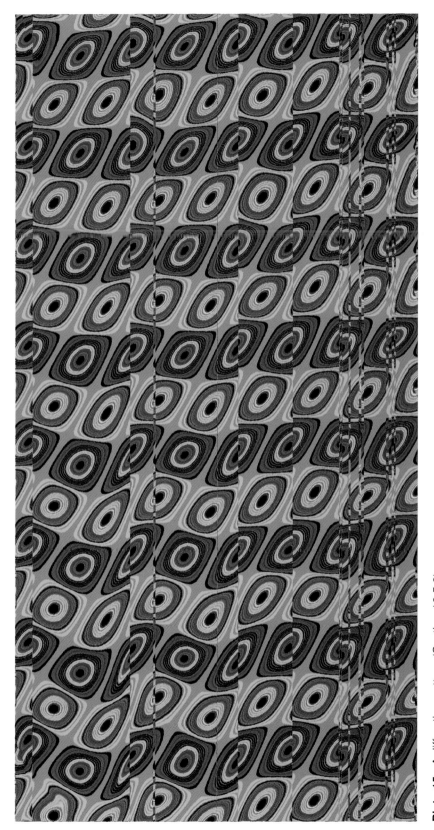

Plate 15: A diffraction pattern (Section 12.5.2).

Plate 16: The Sierpinski sponge. The holes in the ordinary sponge have been filled with spheres. Each level has a distinct appearance (Section 11.4.6).

7.2 Bifurcations

is always unstable (repelling), that is, the orbit of an initial value near the fixed point tends away from the fixed point (because the derivative of f_r at $x = 0$ is larger than 1). For $1 < r < 3$, the second fixed point at $x = (r - 1)/r$ is stable (attracting), that is, the orbits of nearby initial values converge toward the fixed point (see also Plate 1-a). At $r = 3$ this fixed point loses its stability, because the first derivative becomes -1. In this case, convergence toward the fixed point is extremely slow, as the following plot shows. Such fixed points are called *indifferent*.

It is impossible to tell from the plot whether the iteration eventually converges to $x = 2/3$.

```
In[16]:= FunctionIteration[ LogisticsMap[3.0], 0.5, 50 ];
```

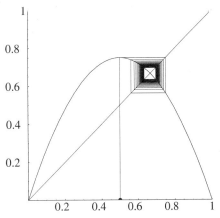

Here is the number of iterations it takes to come within two decimal digits of the fixed point. Convergence to within three digits would take 222, 204 steps.

```
In[17]:= i = 0; \
         FixedPoint[(i++; LogisticsMap[3.0][#])&, 0.5,
            SameTest -> (Abs[#1-#2]<10^-2 &) ]; \
         i
Out[17]= 2210
```

Fixed points can usually be found faster using Newton iteration. `FindRoot[]` uses a variant of Newton iteration internally.

The iterative method converges quickly.

```
In[18]:= FindRoot[ LogisticsMap[3.0][x] == x, {x, 0.5} ]
Out[18]= {x -> 0.666667}
```

Newton iteration can also find unstable fixed points. It is, therefore, not suitable for determining whether ordinary function iteration would converge.

```
In[19]:= FindRoot[ LogisticsMap[3.2][x] == x, {x, 0.5} ]
Out[19]= {x -> 0.6875}
```

The fixed point in this example is repelling, because the absolute value of the derivative is greater than 1.

```
In[20]:= LogisticsMap[3.2]'[x /. %]
Out[20]= -1.2
```

For values $r > 3$, an attractive two-cycle develops. The transition from an attractive fixed point to a two-cycle is an example of a *bifurcation*. It is easily understood if we look at the iterated map $f_r^{(2)}(x) = f_r(f_r(x))$. The operation of repeated function application is easy to code. It is a higher-level function that takes a function f as argument and returns another function, the n-fold iterate or composition $f^{(n)}$.

```
Iteration[n_Integer][f_] := Composition @@ Table[f, {n}]
```

The symbolic result is an expression for $f^{(2)}$.

```
In[21]:= f2 = Iteration[2][f]
Out[21]= Composition[f, f]
```

It can be applied to an argument x.

```
In[22]:= f2[x]
Out[22]= f[f[x]]
```

The reason we look at $f^{(2)}$ is that every two-cycle of f (and every fixed point, too) is a fixed point of $f^{(2)}$. To show the bifurcation, we shall compare the graphical iteration of f_r and $f_r^{(2)}$ for several values of r. We shall use two initial values, chosen so that their iterates end up in different fixed points of $f^{(2)}$ eventually. This code generates the plots for f_r and $f_r^{(2)}$ side by side:

```
figure[r_, x0_, n_:20] :=
    Module[{ga, gb},
        ga = FunctionIteration[ LogisticsMap[r], x0, n,
                DisplayFunction -> Identity, Ticks -> None ];
        gb = FunctionIteration[ Iteration[2][LogisticsMap[r]], x0, n,
                DisplayFunction -> Identity, Ticks -> None ];
        Show[ GraphicsArray[{ga, gb}] ];
    ]
```

Here is the situation shortly before the bifurcation (with $r = 2.8$). All initial points converge toward the one attractive fixed point.

In[23]:= **figure[2.8, {0.5, 0.76}];**

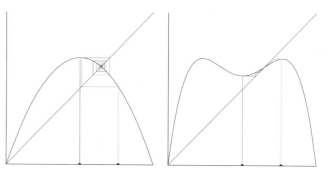

7.2 Bifurcations

At the bifurcation ($r = 3$), convergence is very slow.

In[24]:= `figure[3.0, {0.5, 0.76}];`

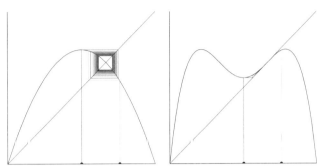

For $r = 3.2$, the central fixed point of $f^{(2)}$ becomes unstable and two new attractive fixed points develop. They correspond to a two-cycle of f itself.

In[25]:= `figure[3.2, {0.67, 0.70}];`

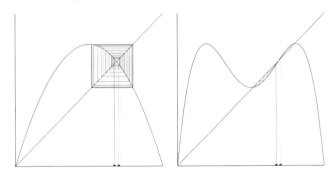

Fixed points and bifurcation points can sometimes be computed exactly.

Here are the two fixed points of the logistics map.

In[26]:= `Solve[ LogisticsMap[r][x] == x, x ]`

Out[26]= `{{x -> 0}, {x -> `$\frac{-1 + r}{r}$`}}`

The first bifurcation occurs when the first derivative at the fixed point becomes -1. We can solve for the corresponding value of r.

In[27]:= `Solve[ LogisticsMap[r]'[(r-1)/r] == -1 ]`

Out[27]= `{{r -> 3}}`

Here are the four fixed points of the second iterate.

In[28]:= `Solve[Iteration[2][LogisticsMap[r]][x] == x, x ] //`
         `        FullSimplify`

Out[28]= `{{x -> 0}, {x -> `$\frac{-1 + r}{r}$`},`

`{x -> `$\frac{1 + r - \sqrt{(-3 + r)(1 + r)}}{2r}$`},`

`{x -> `$\frac{1 + r + \sqrt{(-3 + r)(1 + r)}}{2r}$`}}`

We take one of the new (stable) fixed points to find the next bifurcation point.

```
In[29]:= x /. %[[4]]
          1 + r + Sqrt[(-3 + r) (1 + r)]
Out[29]= ─────────────────────────────────
                        2 r
```

The next bifurcation occurs when the first derivative at this fixed point becomes -1.

```
In[30]:= Solve[ Iteration[2][LogisticsMap[r]]'[%] == -1 ] //
           Simplify
Out[30]= {{r -> 1 - Sqrt[6]}, {r -> 1 + Sqrt[6]}}
```

Here is its numerical value. Only the second solution lies in the interval $1 < r < 4$.

```
In[31]:= N[ % ]
Out[31]= {{r -> -1.44949}, {r -> 3.44949}}
```

A more general method, adapted from [25], finds bifurcation points numerically. We can find fixed points of a function f easily with bisection, using `FindRoot[`f`[x] == x, {x, `x_{min}`, `x_{max}`}]`. To determine whether such a fixed point is attractive, we need to compute the first derivative f' and evaluate it at the fixed point. If the absolute value is less than or equal to 1, we have an attractive fixed point. The bifurcation points are the points where the derivative is equal to 1 or -1. To find higher bifurcation points, we work with $f^{(2^k)}$, for $k = 1, 2, \ldots$. We implemented this method in the auxiliary function `findAttractive[`f`, `f'`, {`x_{min}`, `x_{max}`}, `tol`, `$opts\ldots$`]`. It turns out that `FindRoot[`f`[x] == x, {x, `x_{min}`, `x_{max}`}]` cannot be relied on to find a solution that does indeed lie in the interval (x_{min}, x_{max}). We need to check the result returned carefully. If the fixed point x_m found is not attractive, we search for additional fixed points in the two subintervals (x_{min}, x_m) and (x_m, x_{max}) until no more fixed points are found. The argument tol is used to specify the desired accuracy and to exclude fixed points at the boundaries of the interval (x_{min}, x_{max}). The code of `FindPeriod[]` and `findAttractive[]` is in IteratedFunctions.m, reproduced in Listing 7.2–1.

The function `FindPeriod[`map`, {`x_{min}`, `x_{max}`}, `$opts\ldots$`]` starts with an assumed period of 1 and tries to locate an attractive fixed point. As long as there is no success, the period is doubled and the search is performed again. As soon as it succeeds, it returns the points in the attractive orbit. To avoid building up large expressions, we compute the iterated map and its derivative iteratively, rather than by composition (using `Iteration[p][`map`]`). The function map passed to `findAttractive` is, in fact, an iterative program hidden in the body of a pure function. It looks like `Function[x, Nest[map, x, p]]`. The iteration for the derivative is a bit more complicated. It uses the chain rule according to the formulae

$$\begin{aligned} g^{(0)\prime}(x) &= 1 \\ g^{(k)\prime}(x) &= g'(g^{(k-1)}(x))g'(x). \end{aligned} \qquad (7.2\text{--}1)$$

These formulae lead to the simultaneous iteration

$$\begin{aligned} x_0 &= x \\ x'_0 &= 1 \\ x_{i+1} &= g(x_i) \\ x'_{i+1} &= g'(x_i)x'_i \end{aligned} \qquad (7.2\text{--}2)$$

to compute $g^{(k)\prime}(x) = x'_k$.

7.2 Bifurcations

```
FindPeriod[map_, xr:{_, _}:{0, 1}, opts___?OptionQ] :=
    Module[{p, mapn, mapnp, x, tol, maxp, fp},
        p    = InitialPeriod /. {opts} /. Options[FindPeriod];
        tol  = Tolerance /. {opts} /. Options[FindPeriod];
        maxp = MaxPeriod /. {opts} /. Options[FindPeriod];
        With[{mapp = map'}, fp = Function[{x, xp}, {map[x], mapp[x] xp}] ];
        While[ True,
            (* attractive fixed point *)
            mapn  = Function[ x, Nest[map, x, p] ];
            mapnp = Function[ x, Nest[ fp @@ # &, {x, 1}, p ][[2]] ];
            x = findAttractive[mapn, mapnp, xr, tol, opts];
            If[ x =!= $Failed, Break[] ];
            p *= 2; (* double period *)
            If[ p > maxp, Return[{}]];
        ];
        NestList[map, x, p-1] (* generate orbit *)
    ]
findAttractive[map_, mapp_, {xmin_, xmax_}, tol_, opts___] :=
    Module[{x, xm},
        xm = findFixedPoint[map, {xmin, xmax}, tol, opts];
        If[ xm === $Failed, Return[xm] ]; (* no fixed point *)
        If[ Abs[mapp[xm]] <= 1, Return[xm] ];  (* success *)
        (* left half *)
        x = findAttractive[map, mapp, {xmin, xm}, tol, opts];
        If[ x =!= $Failed && Abs[mapp[x]] <= 1, Return[x] ];
        (* right half *)
        findAttractive[map, mapp, {xm, xmax}, tol, opts]
    ]
findFixedPoint[map_, {xmin_, xmax_}, tol_, opts___] :=
    Module[{x, xx, interval},
        interval = {xmin+2tol, xmax-2tol}; (* open interval *)
        x = xx /. FindRoot[ map[xx] == xx, Evaluate[{xx, interval}],
                AccuracyGoal -> -Log[10.0, tol],
                Evaluate[FilterOptions[FindRoot, opts]] ];
        If[ TrueQ[interval[[1]] < x < interval[[2]] &&
                Abs[x - map[x]] < tol], x, $Failed ]
    ]
```

Listing 7.2–1: Part of IteratedFunctions.m: Finding periods and attractive fixed points

The logistics map $f_{3.5}$ has an attractive periodic orbit of length 4. The default value of FindRoot's option MaxIterations is too small, so we give a larger value.

```
In[32]:= FindPeriod[LogisticsMap[3.5], MaxIterations -> 25]
Out[32]= {0.874997, 0.38282, 0.826941, 0.500884}
```

A simple binary search can now be used to locate bifurcation points. We start with an interval (r_0, r_1) of values for r known to contain a bifurcation point and compute the length of the periodic orbits at the left end ($r = r_0$) and in the middle (at $r_m = (r_0 + r_1)/2$). If the two lengths differ, we know that there is a bifurcation in the left half (r_0, r_m); otherwise, we search in the right half (r_m, r_1). We stop the search as soon as the remaining interval is smaller than some given threshold *delta*. Note that we avoid trying to compute the period at the right end of the interval, because this endpoint may well lie outside the region of period doubling. The code of FindBifurcation[map_r, {r_{min}, r_{max}}, {x_{min}, x_{max}}, *delta*, *opts*...] is shown in Listing 7.2–2.

```
FindBifurcation::usage = "FindBifurcation[map, {r0, r1}, {xmin, xmax}, delta, opts..]
    finds the leftmost bifurcation of map[r] in the interval r0 <= r <= r1 by
    bisection. It stops as soons as the interval is smaller than delta.
    The default x range is {0, 1}."
Options[FindBifurcation] = {
    InitialPeriod -> 1,
    Tolerance -> 1.0 10^-10
};
FindBifurcation[mapr_, {rmin_, rmax_}, xr:{_, _}:{0, 1}, delta_, opts___?OptionQ] :=
    Module[{p, pnew, tol, r0 = rmin, r1 = rmax, rm},
        p = InitialPeriod /. {opts} /. Options[FindBifurcation];
        tol = Tolerance /. {opts} /. Options[FindBifurcation];
        p = Length[FindPeriod[ mapr[r0], xr, InitialPeriod -> p,
                        Tolerance -> tol, opts ]];
        If[ p == 0, Message[FindBifurcation::nocyc, r0]; Return[0] ];
        While[ Abs[r1-r0] > delta,
            rm = (r0+r1)/2;
            pnew = Length[FindPeriod[ mapr[rm], xr, InitialPeriod -> p,
                            Tolerance -> tol, opts ]];
            If[ pnew == 0, Message[FindBifurcation::nocyc, rm]; Return[0] ];
            If[ pnew <= p, r0 = rm, r1 = rm ];
        ];
        (r0+r1)/2
    ]
General::nocyc = "FindPeriod did not converge for r = '1'."
```

Listing 7.2–2: Part of IteratedFunctions.m: Finding bifurcation points

Here is the first bifurcation computed to six decimal places.

```
In[33]:= FindBifurcation[ LogisticsMap, {2.5, 3.6}, 10^-6,
                MaxIterations -> 30 ]
Out[33]= 3.
```

Here is the next one. The option InitialPeriod can be used to speed up the computation if the length of the period at the left end of the search interval is known.

```
In[34]:= FindBifurcation[ LogisticsMap, {%, 3.6}, 10^-6,
                MaxIterations -> 50, InitialPeriod -> 2 ]
Out[34]= 3.44949
```

And one more. Eventually we will lose too much numerical accuracy, however, because of the large number of iterations performed.

```
In[35]:= FindBifurcation[ LogisticsMap, {%, 3.6}, 10^-6,
              MaxIterations -> 50, InitialPeriod -> 4 ]

Out[35]= 3.54409
```

7.3 The Final-State Diagram

The dynamics of a family of functions f_r, such as our logistics map, is best visualized with the final-state diagram. The horizontal axis shows the range of values of r. Above each value of r we plot the limit cycles, if they exist; otherwise, we plot the points reached after a large number of iterations. The chapter-opener picture on page 135 shows a high-resolution rendering of the final-state diagram for the logistics map. Note the bifurcations at $r = 3$, $r \approx 3.45$, and $r \approx 3.54$. Let us see how we can produce such diagrams efficiently.

A naive method to draw a final-state diagram chooses many different values of the dynamics parameter r, then chooses a random initial point x, performs some iterations of f_r to get close to an attractive cycle, should one exist, and then generates a number of points whose horizontal coordinates are r and whose vertical coordinates are the successive iterates. This method works well (and is the only feasible one) for the chaotic region on the right side of the diagram. It performs poorly for the period-doubling region. Many points with essentially identical coordinates are generated unnecessarily.

Because we have the tools to find orbits and calculate periods, we can use them to draw better diagrams. We start with the first value of r and try to find the attractive orbit of low period, if one exists, using `FindPeriod`. If we succeed, we can plot these points (and no others). Eventually the period becomes so high that our numerical methods break down. At this point we start with the general method of performing a large number of iterations.

For the code of `BifurcationDiagram[`map_r`, {`r_0`, `r_1`}, {`x_{min}`, `x_{max}`},` *options*...`]` see Listing 7.3–1. Note that `BifurcationDiagram` takes a map generator as first argument, that is, an object map_r that produces a function when a value of r is applied to it (in the form $map_r[r]$).

There is no general method to produce a good picture for any family map_r of functions. A number of options allow the program to be tweaked. With `PlotPoints`, we can specify the number of different values of r to use. `MaxPoints` gives the number of points to generate for one value of r in the chaotic region. The actual number of points may be smaller. The range of x values actually appearing is used to bound the size of the orbit, leading to smaller graphics objects. The options `InitialPeriod` and `MaxPeriod` tune the behavior of `FindPeriod`. To disable the special code for the period-doubling part of the diagram, use the setting `MaxPeriod -> 0`.

The size of the points drawn is specified with the option `PointSize`. There is a built-in graphic directive `PointSize[s]`. Being a directive, it must be inserted into the list of graphics objects contained in the `Graphics` data. Inserting such a directive into the list is cumbersome.

Therefore, we provide an *option* with the same name. The default is Automatic, which computes the point size from the values of PlotPoints and MaxPoints.

The *sine map* looks similar to the logistics map and, indeed, it shows the same dynamics. Note how we insert the numerical value of π into the body of the pure function to avoid its recomputation every time the function is used.

```
In[36]:= With[{pi = N[Pi]},
            SinMap[r_] := r Sin[# pi] &;
         ]
```

Here is a low-resolution final-state diagram for the sine map. The option MaxIterations is eventually passed to FindRoot.

```
In[37]:= BifurcationDiagram[ SinMap, {0.7, 1},
            PlotPoints -> 300, MaxPoints -> 200,
            MaxIterations -> 25 ];
```

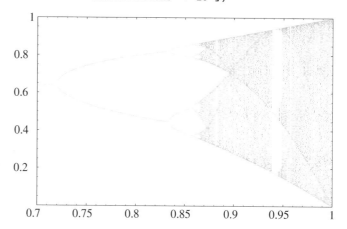

Another option, Levels, can be used to discretize the coordinates of the points to be drawn. The advantage of this method is that it avoids drawing a large number of points with almost identical coordinates, reducing the size of the graphics object and POSTSCRIPT code produced. The disadvantage is that a discretization may introduce visual artifacts not present in the correct diagram. This is especially noticeable at screen resolution if the number of levels chosen is smaller than the number of dots on the screen along the displayed image. An example of this effect is given in the notebook Iterations.nb.

```
BifurcationDiagram::usage = "BifurcationDiagram[map, {r0, r1}, {x0, x1}, opts...]
    draws the bifurcation diagram of map[r] for r0 <= r <= r1.
    The default x range is {0, 1}."

PlotPoints::usage = PlotPoints::usage <>
    " PlotPoints -> n is an option of BifurcationDiagram. It specifies
    the number of different values for r to compute."

MaxPoints::usage = "MaxPoints -> n is an option of BifurcationDiagram
    that gives an upper bound for the number of points in an orbit."
```

7.3 The Final-State Diagram

```
Levels::usage = "Levels -> n is an option of BifurcationDiagram that gives
    the number of discrete values to which points are constrained.
    The default of Infinite performs no quantization."
Options[BifurcationDiagram] = {
    PointSize -> Automatic,
    PlotPoints -> 100,
    MaxPoints -> 200,
    PreIterations -> 200,
    InitialPeriod -> 1,
    MaxPeriod -> 8,
    Levels -> Infinity
};
BifurcationDiagram[ mapr_, {r0_, r1_, dr0_:Automatic}, xr:{x0_, x1_}:{0, 1},
                opts___?OptionQ ] :=
    Module[{points = {}, r = r0, dr = dr0, p, x, cycle,
        map, ps, pp, mp, nmax, levels, pi},
        pp = PlotPoints /. {opts} /. Options[BifurcationDiagram];
        If[ dr === Automatic, dr = N[(r1-r0)/(pp-1)] ];
        pp = Round[(r1-r0)/dr] + 1; (* increment overrides it *)
        mp = MaxPoints /. {opts} /. Options[BifurcationDiagram];
        p = InitialPeriod /. {opts} /. Options[BifurcationDiagram];
        maxp = MaxPeriod /. {opts} /. Options[BifurcationDiagram];
        maxp = Min[maxp, mp];
        levels = Levels /. {opts} /. Options[BifurcationDiagram];
        pi = PreIterations /. {opts} /. Options[BifurcationDiagram];
        ps = PointSize /. {opts} /. Options[BifurcationDiagram];
        If[ ps === Automatic,
            ps = Min[(x1-x0)/(r1-r0)/Min[2levels, mp]/2.0, (r1-r0)/pp/2.0]
        ];
        x = Random[Real, xr];

        While[ p <= maxp && r <= r1,
            map = mapr[r];
            cycle = FindPeriod[map, xr, InitialPeriod -> p, MaxPeriod -> maxp, opts];
            p = Length[cycle];
            If[ p == 0, Break[] ];
            x = First[cycle];
            If[ x == x0 || x == x1, x = Random[Real, xr] ];
            cycle = {r, #}& /@ cycle;
            AppendTo[points, cycle];
            r += dr;
        ];
        (* chaotic rest *)
        map = mapr[r];
        x = Nest[map, Random[Real, xr], pi];
        cycle = NestList[ map, x, mp ];
        {min, max} = Through[{Min, Max}[cycle]]; (* range predictor *)
        Do[ map = mapr[r];
            x = Nest[map, Random[Real, xr], pi];
            p = Ceiling[ mp (max-min)/(x1 - x0) ];
```

```
            cycle = NestList[ map, x, p];
            {min, max} = Through[{Min, Max}[cycle]];
            If[ levels < Infinity, cycle = Discretize[cycle, xr, levels] ];
            cycle = {r, #}& /@ cycle;
            AppendTo[points, cycle];
          ,{r, r, r1, dr}
        ];
        points = {PointSize[ps], Map[Point, points, {2}]};
        Show[ Graphics[ points,
                    {FilterOptions[Graphics, opts],
                     Frame -> True, PlotRange -> {{r0, r1}, xr}} ]
        ]
    ]
Discretize[vals_List, {x0_, x1_}, n_] :=
    Module[{oc},
        oc = Table[0, {n+1}];
        Scan[(oc[[Floor[(#-x0)/(x1-x0) n] + 1]] = 1)&, vals];
        x0 + (Flatten[Position[oc, 1]]-0.5)/n*(x1-x0)
    ]
```

Listing 7.3–1: Part of IteratedFunctions.m: Final-state diagrams

The period-doubling region of the final-state diagram shows that the distance between bifurcations becomes smaller and smaller. Feigenbaum observed that the ratio of successive differences approaches a limit that is independent of the function being iterated [19]. This limit is called the *Feigenbaum constant*. Our crude method for locating the first few bifurcations does not give us enough data to compute this constant to a sufficient accuracy. We show a better way to compute it in Section 7.5. The geometric convergence of the differences of successive bifurcations means also that the sequence of bifurcation points itself has a limit, the *Feigenbaum point*. For the logistics map, it is at $r = 3.569945\ldots$. It is the beginning of the chaotic region of the diagram, following the period-doubling region.

7.4 The Ingredients of Chaos

A function f is chaotic if the orbits of its iterates show these three properties:

- sensitivity to initial conditions
- mixing
- dense periodic orbits.

7.4 The Ingredients of Chaos

We shall explain these properties, as well as two related ones—ergodic orbits and intermittency—with one of the simplest examples of chaotic functions, the logistics map f_r. We shall see that this map is chaotic for values of r in the range from $r \approx 3.5$ to $r = 4$. We have already seen the "route to chaos," that is, the period-doubling region up to the Feigenbaum point. Let us now investigate what happens beyond this point. Rigorous proofs of many of the chaotic properties of the logistics map and other functions can be found in [46]. Our exposition follows this book closely. It is a rich source of examples.

7.4.1 Sensitivity to Initial Conditions

The first characteristic of chaos is sensitivity to initial conditions. Sensitivity means that small differences in initial conditions are magnified arbitrarily large. We can investigate it by observing the long-term behavior of nearby initial values. We have already seen an example of graphical iteration of a number of close initial points (see page 141); another illustration with 500 initial points is reproduced in Plate 1-b.

Let us measure the amount of dispersion of nearby initial points, having distance δ. The amplification of "error" after n iterations is given by

$$\frac{\left| f^{(n)}(x_0 - \frac{\delta}{2}) - f^{(n)}(x_0 + \frac{\delta}{2}) \right|}{\delta}, \qquad (7.4\text{--}1)$$

that is, the final distance of the points divided by their initial distance. We want to measure this quantity for $n = 1, 2, \ldots$.

Our example is the logistics map f_4.	`In[1]:= f4 = LogisticsMap[4]` `Out[1]= 4 #1 (1 - #1) &`
After 30 iterations, the orbits of the two nearby points have moved far apart.	`In[2]:= Nest[f4, {0.4, 0.400000001}, 30]` `Out[2]= {0.784578, 0.668791}`

This command measures the error amplification according to Equation 7.4–1.	`In[3]:= amplification[map_, x0_, delta_, n_] :=`
	`        Abs[ Subtract @@`
	`             Nest[map, {x0-delta/2, x0+delta/2},n]`
	`           ] / delta`

Here are the results for an initial error $\delta = 10^{-6}$ over 20 iterations.	`In[4]:= Table[ amplification[f4, 0.4, 10^-6, i], {i, 20} ]`
	`Out[4]= {0.8, 2.944, 8.15841, 1.30718, 5.21196, 20.5807,`
	`         78.1296, 250.469, 285.194, 955.899, 1545.8, 4162.01,`
	`         1562.13, 6138.28, 22835.7, 66686.5, 17883.6, 70255.1,`
	`         261277., 772041.}`

This logarithmic plot shows that the error roughly doubles at each iteration. A more precise statement is that the first Liapunov exponent of f_4 is log 2.

In[5]:= `ListPlot[ Log[%], PlotStyle -> PointSize[0.01] ];`

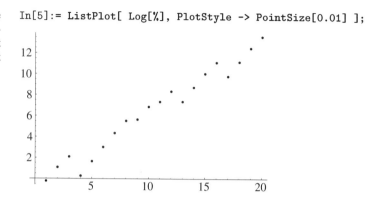

7.4.2 Mixing

The mixing property means that initial points chosen from a small interval will eventually reach any other given interval. In this numerical experiment, we choose 10,000 initial points in the interval $(0.2 \pm 10^{-10}/2)$ and count the number of orbits that do not land in the target interval $(0.68, 0.69)$. More and more orbits will eventually hit the target.

We generate the initial points.

In[6]:= `vals = 10^4; \`
`size = 10.0^-10; \`
`xs = 0.2 + size Range[-vals/2, vals/2]/vals;`

This function chooses all points that do not lie in the target interval.

In[7]:= `survivors[points_List, {x0_, x1_}] :=`
`Select[points, !(x0 < # < x1)&]`

We iterate 1000 times and count the number of survivors after each iteration. Only the survivors are iterated further.

In[8]:= `Table[`
`Length[xs = survivors[f4 /@ xs, {0.68, 0.69}]],`
`{1000} ];`

The logarithmic plot of the number of survivors shows an exponential decay. Such "power laws" are typical of chaotic phenomena.

In[9]:= `ListPlot[ Log[10.0, %] ];`

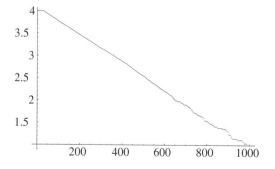

7.4 The Ingredients of Chaos

7.4.3 Dense Periodic Orbits

The third characteristic of chaos is the existence of dense periodic orbits. Given any point, another point belonging to a periodic orbit can always be found arbitrarily close. These periodic orbits are not attractive, however, so any numerical computation will eventually drift away from the orbits. They do not show up in the final-state diagram.

The initial value $\sin^2(\pi/7)$ generates an orbit with period 3. See [46] for a method to find more periodic points.

```
In[10]:= c3 = N[Sin[Pi/7]^2]
Out[10]= 0.188255
```

We can see that after 50 iterations the points drift slowly away from the periodic orbit.

```
In[11]:= FunctionIteration[ f4, c3, 50 ];
```

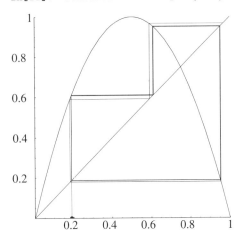

7.4.4 Ergodic Orbits

An *ergodic orbit* is an orbit that comes arbitrarily close to any point in the unit interval. In a numerical experiment we count how often an orbit visits small intervals using functions from the standard packages.

We need these packages.

```
In[12]:= Needs["Statistics`DataManipulation`"];\
         Needs["Graphics`Graphics`"]
```

We generate 10^4 points of the orbit of 0.4.

```
In[13]:= NestList[ f4, 0.4, 10^4 ];
```

We divide the unit interval into 1000 subintervals and count how often each subinterval is visited.

```
In[14]:= BinCounts[ %, {0, 1, 10^-3} ];
```

Here is a bar chart of the result.   `In[15]:= BarChart[ %, BarLabels -> None, BarEdges -> None ];`

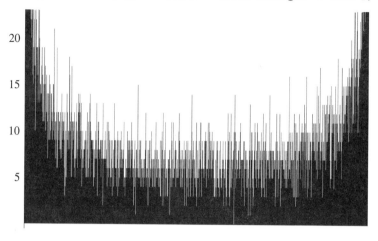

Only one of the subintervals was not hit.

`In[16]:= Count[ %%, 0 ]`
`Out[16]= 1`

7.4.5 Intermittency

Finally, a look at intermittency. It can occur in connection with tangent fixed points, that is, points at which the line $y = x$ is tangent to the graph of f.

For $r = 1 + \sqrt{8}$, there is a tangent fixed point of $f_r^{(3)}$, as seen in this plot. This tangent fixed point of $f_r^{(3)}$ leads to a three-cycle of f_r itself.

```
In[17]:= FunctionIteration[
            Iteration[3][LogisticsMap[N[1 + Sqrt[8]]]],
            0.5, 10
         ];
```

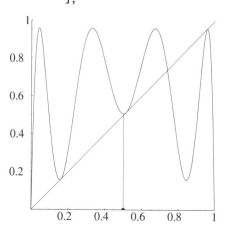

7.5 Super-Attractive Orbits 157

For values of r slightly smaller than $1 + \sqrt{8}$ there is no fixed point, but a small "tunnel" between the graph of $f_r^{(3)}$ and the line $y = x$. Points near the former fixed point (which is at $x = 0.514298$) can spend a long time iterating through this tunnel. Eventually they escape but are thrown back into this region sometime later.

```
In[18]:= FunctionIteration[
            Iteration[3][LogisticsMap[3.8282]],
            0.5, 27 ];
```

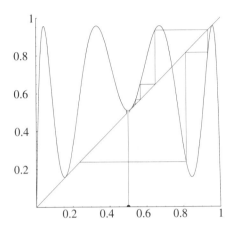

The iterations in this tunnel correspond to near three-cycles of f_r, clearly visible in the time series. Theses cycles are interrupted by short bursts of chaotic iterations. Such behavior is called intermittency.

```
In[19]:= TimeSeries[ LogisticsMap[3.8282], 0.5, 300,
            AspectRatio -> 1/2 ];
```

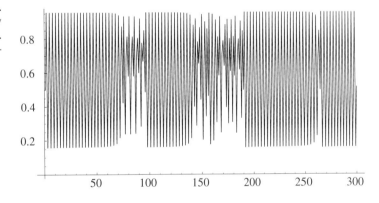

7.5 Super-Attractive Orbits

Within a range of values of r that have attractive orbits of the same length (that is, between two bifurcations), there is one value for which the convergence toward the attractive orbit is especially fast. Such an orbit is called *super attractive*. A super-attractive orbit occurs if the critical point (where the function has its local maximum) is a member of the orbit. For the logistics map, the critical point is $x_{crit} = 1/2$. The first few values of r with super-attractive orbits can be found exactly. To find such an orbit of length n, we solve the equation

$$f_r^{(n)}(x_{crit}) = x_{crit} \qquad (7.5\text{--}1)$$

for r. Remember that a fixed point of $f_r^{(n)}$ corresponds to a cycle of length n for f_r.

Here is the first super-attractive value for an orbit of length 1, that is, a fixed point.

```
In[20]:= Solve[ LogisticsMap[r][1/2] == 1/2, r ][[1]]
Out[20]= {r -> 2}
```

This graphical iteration shows the quick convergence toward the fixed point.

```
In[21]:= FunctionIteration[ LogisticsMap[2], 0.2, 10 ];
```

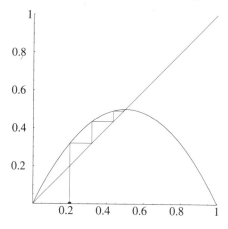

The solutions for $f^{(2)}$ give us the next super-attractive value (together with the first one and a solution out of range).

```
In[22]:= Solve[Iteration[2][LogisticsMap[r]][1/2] == 1/2, r]
Out[22]= {{r -> 2}, {r -> 1 - Sqrt[5]}, {r -> 1 + Sqrt[5]}}
```

Higher super-attractive values can no longer be found in closed form, because Equation 7.5–1 has degree $2^n - 1$ in r and no general formula for the solution of equations of degree higher than 4 exists.

An efficient numerical method to find higher-order super-attractive values has been described in [46]. The idea is to solve Equation 7.5–1 using Newton iteration. Let $g(r) = f_r^{(m)}(x_{crit}) - x_{crit}$, for some $m \geq 0$. A super-attractive value is a zero of g. Given an estimate r_0 of the super-attractive value for r, a better estimate is computed using Newton's formula:

$$r_{k+1} = r_k - \frac{g(r_k)}{g'(r_k)} \qquad (7.5\text{--}2)$$

where g' denotes differentiation with respect to r. Values of $g(r)$ should again be computed by iterating f_r (compare with Equation 7.2–2). This time, however, differentiation is with respect

7.5 Super-Attractive Orbits

to r. For a given value of r, the simultaneous iteration

$$\begin{aligned}
x_0 &= x_{crit} \\
x'_0 &= 0 \\
x_{i+1} &= f_r(x_i) \\
x'_{i+1} &= \frac{d}{dr} f_r(x)\Big|_{x=x_i,\, \frac{dx}{dr}=x'_i}
\end{aligned} \qquad (7.5\text{--}3)$$

gives $g(r) = x_m - x_0$ and $g'(r) = x'_m$.

We assume that we know good initial estimates for s_1, s_2, and s_3, the first three super-attractive values. It can be shown that the ratio of successive differences,

$$\delta_n = \frac{s_{n-1} - s_{n-2}}{s_n - s_{n-1}}, \qquad (7.5\text{--}4)$$

converges toward the Feigenbaum constant, as does the ratio of successive differences of the bifurcation points. Therefore, we can use Equation 7.5–4 to find estimates for higher s_n. If only two initial values are available, we can use an estimate $\delta_3 = 4$. The method converges quickly; only a few Newton iterations are necessary. Almost all time in the algorithm is taken to compute $g(r)$ and $g'(r)$ according to Equation 7.5–3. The number of steps necessary to find s_n is $m = 2^n$.

The code is in the command SuperAttractiveSeries[map_r, x_{crit}, s_{init}, n_{max}], also in the package IteratedFunctions.m (Listing 7.5–1). If we want to compute the sequence s_n to high precision, we need to set $MinPrecision to the value of the desired working precision. The large number of iterations performed would quickly lower the precision of the number involved, but Newton iteration allows us to correct the errors; therefore, we can work with a fixed number of digits. Note the use of Block to give $MinPrecision its special value locally, without disturbing its (global) value. Note how we construct the pure function for the iteration to compute the derivative (Equation 7.5–3). The form Function @@ {vars, body} allows the body to evaluate before the function is constructed. We do this so that the differentiation and simplification are performed at definition time, rather than every time the function is used later. The rules {x -> xi, Dt[x, r] -> xpi} make sure that the differentiation is performed first, before values are inserted. This corresponds to Equation 7.5–3.

Another programming idea worth mentioning is the implementation of a simultaneous iteration. Given the iteration $x_{i+1} = f(x_i, y_i)$, $y_{i+1} = g(x_i, y_i)$, we can construct a pure function taking x_i and y_i as arguments and returning the pair $\{x_{i+1}, y_{i+1}\}$:

    step = Function[{x, y}, {f[x, y], g[x, y]}] .

To compute iterates, we apply this function repeatedly to the initial values $\{x_0, y_0\}$, using Nest. The straightforward command Nest[step, {x_0, y_0}, n] does not work, however, because Nest passes the pair of values as a *single* argument to step. We need to use Apply (infix @@), constructing another pure function:

    Nest[step @@ # &, {x_0, y_0}, n] .

```
SuperAttractiveSeries::usage = "SuperAttractiveSeries[map, xcrit, sinit, nmax]
    computes nmax super-attractive values of r for map[r].
    At least two initial values must be given in the list sinit."
DeltaEstimates::usage = "DeltaEstimates[s] computes successive approximation
    of Feigenbaum's constant from the list s of super-attractive values."

SuperAttractiveSeries[mapr_, xcrit_, sinit_List, nmax_] /; Length[sinit] > 1 :=
    Module[{s = sinit, r, workingprec = Precision[sinit], unit,
            n, delta, s0, iter, gnest, gr, grp, corr},
        DefineIterator[rr_] :=
            Module[ {xi, xpi, x, r},
                Function @@ {{xi, xpi}, Simplify[{mapr[r][xi],
                    Dt[mapr[r][x], r] /. {x -> xi, Dt[x, r] -> xpi}} /. r -> rr}] ];
        unit = 10.0^-(workingprec-1);
        Block[{$MinPrecision = workingprec},
            While[Length[s] < nmax, n = Length[s];
                If[ n == 2, delta = 4, delta = (s[[-2]] - s[[-3]])/(s[[-1]] - s[[-2]]) ];
                s0 = s[[-1]] + (s[[-1]] - s[[-2]])/delta;
                iter = 2^n;
                While[True,
                    gnest = DefineIterator[s0];
                    {gr, grp} = Nest[gnest @@ # &, {xcrit, 0}, iter] - {xcrit, 0};
                    corr = gr/grp;
                    If[ Abs[corr/s0] < unit, Break[] ]; (* converged *)
                    s0 -= corr ];
                AppendTo[s, s0];
            ]];
        s ]

DeltaEstimates[s_List/; Length[s] > 2] :=
    With[{sn = Take[s, {3, -1}], sn1 = Take[s, {2, -2}], sn2 = Take[s, {1, -3}]},
        (sn1 - sn2)/(sn - sn1) ]
```

Listing 7.5–1: Part of IteratedFunctions.m: Computing super-attractive points

An alternative is to define `step` to take a list as a single argument. We can use pattern matching:

$$\text{step}[\{x\_, y\_\}] := \{f[x, y], g[x, y]\}.$$

Solving $f_r^{(4)}(1/2) = 1/2$ numerically should give us three initial super-attractive values. We compute them to 30 digits. There are fifteen solutions, some of which are complex or out of range.

```
In[23]:= NSolve[ Iteration[4][LogisticsMap[r]][1/2] == 1/2,
            r, 30 ] // Short
Out[23]//Short=
  {{r -> -1.96027012722115260157125799}, {r -> -1.4..82},
   <<12>>, {r -> 3.96027012722115260157126}}
```

We filter out the positive real values in order.

```
In[24]:= Cases[ r /. %, _Real?Positive ] // Sort
Out[24]= {2.000000000000000000000000,
         3.236067977499789696409017, 3.498561699327701519999895,
         3.960270127221152601571260}
```

7.5 Super-Attractive Orbits

We need the first three values to start the iteration (the last value does not belong to the period-doubling region).

```
In[25]:= Take[ %, 3 ]
Out[25]= {2.000000000000000000000000,
          3.236067977499789696440917, 3.498561699327701519999895}
```

We compute the first 16 super-attractive values. The computation takes about 270 seconds on a SPARCstation 20.

```
In[26]:= SuperAttractiveSeries[LogisticsMap, 1/2, %, 16] //
          TableForm
Out[26]//TableForm= 2.000000000000000000000000
                    3.236067977499789696440917
                    3.498561699327701519999895
                    3.554640962769921865366608
                    3.566667379856268513977263
                    3.569243531637110337808250
                    3.569795293749944620515350
                    3.569913465422348514840970
                    3.569938774233305487793450
                    3.569944194608064933243630
                    3.569945355486468580892800
                    3.569945604111078438134120
                    3.569945657358856499729630
                    3.569945668762899968347100
                    3.569945671205298854528910
                    3.569945671728383474205070
```

Equation 7.5–4 gives us a way to approximate the Feigenbaum constant. The last result is correct to 10 digits.

```
In[27]:= DeltaEstimates[ % ] // TableForm
Out[27]//TableForm= 4.708943013540503313177
                    4.680770998010695381631
                    4.662959611114410258401
                    4.668403925918400237800
                    4.668953740967622781000
                    4.669157181328843482000
                    4.669191002485096150000
                    4.669199470547725770000
                    4.669201134601042200000
                    4.669201509513552000000
                    4.669201587522380000000
                    4.669201604512200000000
                    4.669201608115900000000
                    4.669201608892000000000
```

7.6 Conclusions

We have not investigated all phenomena of chaos in the logistics map. A closer look at the final-state diagram on page 135 should give you enough ideas for many more investigations. Note especially the region with an attractive periodic orbit of length 3 appearing suddenly out of chaos near $r = 3.832$. For an advanced mathematical treatment of chaotic dynamics, see [16].

Apart from presenting this fascinating subject, I wanted to show that a wide range of visualizations can easily be developed in *Mathematica*. In fact, if you look at the BASIC or C programs often used, you will observe that most of the code deals with graphics issues and that symbolic or higher numerical methods are never used. It is still nontrivial, however, to develop a coherent set of *Mathematica* programs for visualizations. The common ideas that are independent of the subject area being visualized should be identified and programmed only once. For example, a general principle, adopted here, is that any command that takes a single object as its argument should also be able to deal with a list of such objects. Different visual clues (color in our case) are automatically used to distinguish the different objects. Our collection of visualizations for mathematics [23, 24] is based on such principles.

The CD-ROM contains the package IteratedFunctions.m, as well as the notebook Iterations.nb, with the examples from this chapter.

Chapter Eight

Fractional Brownian Motion

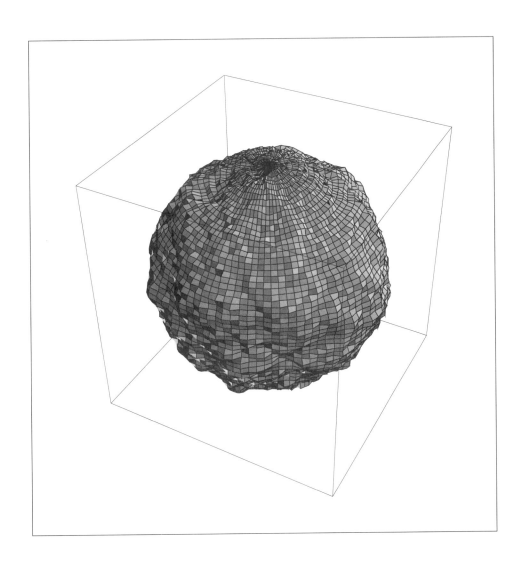

Fractional Brownian motion is a generalization of ordinary Brownian motion that has been used successfully to model a variety of natural phenomena, such as terrains, coastlines, and clouds. We shall briefly describe the theory behind fractional Brownian motion and then look at three methods to generate such data in *Mathematica*. These methods pose a number of interesting programming problems. We shall also develop methods to visualize fractional Brownian motion in *Mathematica* and with external renderers.

About the illustration overleaf:

A planet or asteroid with a fractal surface. The surface is the result of 2000 random spherical faults applied to a sphere (see Section 8.4.2).

The code for this picture is in BookPictures.m.

8.1 Introduction

Brownian motion is a process in which a particle is subjected to random displacements. In a simple one-dimensional model the displacements are either $+1$ or -1, chosen uniformly. The position of the particle $V(t)$ at time t is found by adding the random displacements

$$V(t) = \sum_{i=0}^{t} A_i$$

where A_i is a random variable taking on the values $+1$ and -1 with equal probability. Our main interest is in the *trace* of the particle, that is, a plot of $V(t)$ versus t.

This random variable takes on the values $+1$ and -1 with equal probability.

```
In[1]:= randDir := 2 Random[Integer] - 1
```

This function generates a random walk of length n by successively adding random displacements.

```
In[2]:= randWalk[n_] := NestList[# + randDir&, 0, n-1]
```

We generate a random walk of length 1000.

```
In[3]:= (walk = randWalk[1000]) // Short
Out[3]//Short=
 {0, 1, 0, 1, 2, 1, 2, 1, 2, 3, 4, 5, 6, 7, 6, 5, 4, 3, 2,
   <<972>>, 11, 10, 11, 12, 11, 12, 13, 12, 11}
```

Here is the corresponding plot of the displacements.

```
In[4]:= ListPlot[%, PlotJoined -> True, AspectRatio -> 1/3];
```

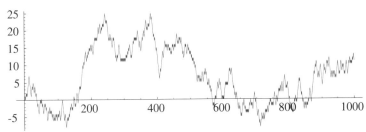

One characteristic of such random walks is that the average displacement is proportional to the square root of the time difference, that is

$$|V(t) - V(t+dt)| \propto dt^{0.5}. \tag{8.1–1}$$

Here is the average displacement with $dt = 2$. The value 2.0 (instead of 2) makes sure that the result is a floating-point number rather than a rational number.

```
In[5]:= Plus @@ (Abs[Drop[walk, -2] - Drop[walk, 2]])/
         (Length[walk] - 2.0)
Out[5]= 1.00802
```

The average displacement with $dt = 4$ is about $\sqrt{2} = 1.41$ times larger. The observed value may differ for such small sample sizes.

```
In[6]:= Plus @@ (Abs[Drop[walk, -4] - Drop[walk, 4]])/
              (Length[walk] - 4.0)
Out[6]= 1.51606
```

8.1.1 Fractional Brownian Motion

Fractional Brownian motion (fBm) is a generalization of ordinary Brownian motion with the scaling property

$$|V(t) - V(t+dt)| \propto dt^h \tag{8.1-2}$$

for arbitrary h, $0 < h < 1$. The exponent h is called the *Hurst exponent*. Ordinary Brownian motion (Equation 8.1–1) has $h = 1/2$.

The random walks we want to study are finite approximations of a limit curve obtained as the number of displacements tends to infinity and the distance between successive displacements tends to zero. The limit curve is a statistically self-similar fractal. Because of the randomness involved it is not exactly self-similar as is an ordinary fractal (as described in the first volume of this book, Chapter 8). The scaling property of Equation 8.1–2 is satisfied only on average.

The concept of fractional Brownian motion can be generalized to higher dimensions: the variable t becomes a vector. If t is two-dimensional, the function $V(t)$ describes a surface in space. In the discrete case, the values of V are given in a matrix. We make use of the three techniques provided by *Mathematica* to visualize such data: surface graphics, contour graphics, and density graphics.

We develop three different methods to generate fBm data. The functions that generate the data will have this calling structure

$$\texttt{fBm}Method[e, n, h, parameters]$$

where e is the dimension of the space, n is the grid size, and h is the Hurst exponent. Some of the methods take additional optional parameters. The result is an $n \times n \times \cdots \times n$ tensor of rank e. For $e = 1$ we get a simple list of values, and $e = 2$ gives us a matrix. Applications with higher dimensions are rare. The first of our methods is iterative. It allows the refinement of existing lower-resolution fBm data. If this is desired, the old data can be given as the first parameter, instead of e.

8.1.2 Random Variables

The main tool for introducing randomness is a Gaussian random variable, that is, a random variable with a normal distribution and variance 1. Such random variables are available in the standard package Statistics'NormalDistribution'. Listing 8.1–1 displays the code from our package fBm.m that defines the necessary random variables.

8.2 Random Additions

```
Needs["Statistics`NormalDistribution`"]
With[{nd = NormalDistribution[], pi2 = N[2Pi]},
  gauss := Random[nd];
  randPhase := Exp[I Random[Real, {0, pi2}]];
  rand3D := With[{phi = Random[Real, {0, pi2}],
              z = Random[Real, {-1, 1}]},
            {Sqrt[1-z^2] Cos[phi], Sqrt[1-z^2] Sin[phi], z}]
]
```

Listing 8.1–1: Part of fBm.m: Random variables

We read in the package with all functions described in this article.	`In[7]:= Needs["MathProg`fBm`"]`
gauss gives normally distributed random values.	`In[8]:= Table[gauss, {5}]` `Out[8]= {1.33414, 0.602033, 1.23856, 0.797209, -1.54829}`
randPhase gives a random point on the unit circle, expressed as a complex number, which we will need in Section 8.3.	`In[9]:= randPhase` `Out[9]= -0.0803974 + 0.996763 I`
rand3D gives random points on the unit sphere, which we will need in Section 8.4.	`In[10]:= rand3D` `Out[10]= {-0.0597678, 0.827735, -0.557927}`

8.2 Random Additions

The first of our methods is iterative. It is useful to picture the data tensor as an $n \times n \times \cdots \times n$ grid of points with values attached to them. Our method starts with an initial grid of $2 \times 2 \times \cdots \times 2$ points, with values chosen with a Gaussian distribution. Each iteration refines a given grid by adding new points with values given by linear interpolation, and then changing the values at all points (old and new) by a random amount. The variance of the random displacements can be chosen to achieve any desired Hurst exponent. This iterative process is called *random addition*.

The formulae for choosing the variance are given in Peitgen and Saupe's book [47]. Let us concentrate on the programs instead. A single step, `fBmRAStep[`*grid*`, `h`, `r`]`, takes an existing grid and computes a new grid with $1/r$ times as many points. The parameter r, $0 < r < 1$, is the *lacunarity*. Most often, we will use $r = 1/2$ to double the number of points. In the one-dimensional case we can simply perform linear interpolation with the built-in function `Interpolation[`*data*`, InterpolationOrder->1]`. The interpolation points are $1, 2, \ldots n$, where n is the length of the list *data*. To obtain $m = n/r$ new points, we evaluate the resulting interpolating function at points $1, \frac{n-1}{m-1}, 2\frac{n-1}{m-1}, \ldots, n$.

In this small example, we use three points and want to interpolate to five points.

```
In[1]:= data = {3., 2., 1.}; n = Length[data]; m = 5;
```

We construct the interpolation function.

```
In[2]:= if = Interpolation[data, InterpolationOrder -> 1]
Out[2]= InterpolatingFunction[{{1., 3.}}, <>]
```

Now, we can evaluate it at the new points. The interpolating function is listable, so we can give it the whole list of new points as the argument.

```
In[3]:= if[ Range[1, n, (n-1)/(m-1)] ]
Out[3]= {3., 2.5, 2., 1.5, 1.}
```

Here is the code that performs these steps:

```
inter[x_, m_] :=
    With[{interpol = Interpolation[x, InterpolationOrder->1], n = Length[x]},
        interpol[ Range[1, n, (n-1)/(m-1)] ]
    ]
```

As explained in Section 8.1.1, the function fBmRA can be called in several ways. In the form fBmRA[e, n, h, r] it is used to generate an e-dimensional tensor, with each dimension of size n. In the form fBmRA[*data*, n, h, r] the given tensor *data* is refined by as many random-addition steps as are necessary to extend it to size n. The fractal dimension of the resulting fBm data is $d = e + 1 - h$. For convenience, we allow the data also in the form of a surface, contour, or density graphic. The first element of these graphic data structures is a matrix, which we extract, subject to the random additions, and then put back into the data structure. This code is shown in Listing 8.2–1.

```
fBmRA[e_Integer, args__] := fBmRA[ Array[gauss/2&, Table[2, {e}]], args ]
fBmRA[mat_List, n_Integer, h_, r_:1/2] :=
    Module[{x = mat},
        While[ Length[x] < n, x = fBmRAStep[x, h, Max[r, Length[x]/n]] ];
        x
    ]
fBmRA[(g:(SurfaceGraphics|ContourGraphics|DensityGraphics))[mat_, opts___], args__] :=
    g[ fBmRA[mat, args], opts ]
```

Listing 8.2–1: Part of fBm.m: fBm with random additions

Here is a sample one-dimensional random walk of length 16 with $h = 0.5$.

```
In[4]:= fBmRA[1, 16, 0.5]
Out[4]= {1.22422, 1.35711, 1.66076, 1.20886, 0.955561,
    0.816783, 0.349294, -0.0667059, 0.3083, 0.135712,
    -0.125079, -0.0251158, 0.21223, 0.192075, 0.571294,
    0.629977}
```

8.2 Random Additions

In this table, we generate fBms consisting of 1024 points each and let the Hurst exponent vary from 0.8 down to 0.2.

```
In[5]:= plots =
        Table[ ListPlot[fBmRA[1, 1024, h], Axes -> None,
            PlotJoined -> True, DisplayFunction -> Identity,
            PlotRange -> All, AspectRatio -> 0.3,
            PlotLabel -> StringForm["h = `1`", h]],
          {h, 0.8, 0.2, -0.2} ];
```

Here are the resulting fractal curves.

```
In[6]:= Show[ GraphicsArray[Partition[plots, 2]],
          GraphicsSpacing -> {0.1, 0.3}
        ];
```

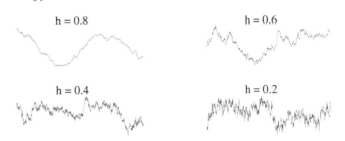

The fractal dimension of an fBm curve is equal to $2 - h$. Therefore, the curves with smaller h look much wilder than the curves with larger h.

If the dimension e is greater than 1, we need to perform multidimensional linear interpolation. Fortunately, this interpolation is, well, *linear*, so we can do it coordinate by coordinate. Expressing the method in a way that is independent of the dimension, and without loops, is an interesting programming challenge. Recall that a tensor of rank e is stored as a nested list, with e levels of nesting. The points belonging to the last dimension are, therefore, stored neatly as lists at level $e - 1$. We can simply map the one-dimensional interpolation method inter discussed earlier onto all these lists. Interpolating the other levels is not that easy. The elements are not stored in adjacent positions. But we can transpose our tensor to get any desired dimension to the end, perform the interpolation, and then transpose back. We can avoid the intermediate transpositions by choosing a cyclic transposition that brings one level after the other to the end and, after e applications, puts the tensor back into the original order. Such a transposition is given by Transpose[*tensor*, {e, 1, 2, ..., $e - 1$}]. The required permutation is most easily obtained as RotateRight[Range[e]].

Here is a sample tensor of rank 2.

```
In[7]:= {{0,1}, {2, 4}};
```

We interpolate the first dimension to a new size 3.

```
In[8]:= Map[ Function[slice, inter[slice, 3]],
          Transpose[%, {2, 1}],
          {1}
        ]

Out[8]= {{0, 1, 2}, {1, 5/2, 4}}
```

A second, identical, step interpolates the second dimension and returns the tensor to its original order.

```
In[9]:= Map[ Function[slice, inter[slice, 3]],
           Transpose[%, {2, 1}],
           {1} ]

Out[9]= {{0, 1/2, 1}, {1, 7/4, 5/2}, {2, 3, 4}}
```

We can use Fold to interpolate each dimension to a given new size. The desired sizes are given in a list. Listing 8.2–2 shows the interpolation code and the definition of fBmRAStep[*tensor*, h, r], which performs one random-addition step.

```
interpolate[x_, ndims_List] :=
  With[{e = TensorRank[x]},
    With[{trans = RotateRight[Range[e]]},
      Fold[ Map[ Function[slice, inter[slice, #2]],
               Transpose[#1, trans], {e-1} ]&,
           x, ndims ]
    ]
  ]

fBmRAStep[mat_List, h_, r_:1/2] :=
  Module[{t, T, odims, ndims},
      odims = Dimensions[mat]; ndims = Ceiling[odims/r];
      t = 1.0/(Max[ndims]-1); T = 1.0/(Max[odims]-1);
      With[{delta = Sqrt[0.5] Sqrt[1 - (t/T)^(2-2h)] t^h},
          Map[delta gauss + # &, interpolate[mat, ndims], {-1}]
      ]
  ]
```

Listing 8.2–2: fBm.m: Interpolation and random-addition steps

The values t and T are the reciprocals of the new and old number of grid points. They are needed to compute the size of the random additions. The new dimensions are equal to the old dimensions divided by r.

Here is a simple example for a two-dimensional tensor. The number of points is doubled and random displacements with a distribution leading to a Hurst exponent of 1/2 are added.

```
In[10]:= fBmRAStep[{{0,1}, {2, 4}}, 0.5, 1/2] // MatrixForm
Out[10]//MatrixForm=
  0.530629   0.369684   1.3835    1.60844
  0.263218   0.823008   1.28386   2.39651
  1.15738    1.79442    2.15509   3.10854
  1.55899    2.80669    2.81409   4.04741
```

Here are some examples of fBm data and their visualization in two and three dimensions.

8.2.1 Landscapes

Landscapes can be generated as two-dimensional fBm data and are easily visualized as surface or contour graphics.

8.2 Random Additions

We generate the data for a surface with fractal dimension $3 - h = 2.25$. It consists of 64×64 points.

`In[11]:= landscape0 = fBmRA[2, 64, 0.75];`

Here is the matrix, rendered as a surface graphic. For such large numbers of plot points, it is best to turn off the mesh.

`In[12]:= Show[SurfaceGraphics[landscape0], Mesh -> False];`

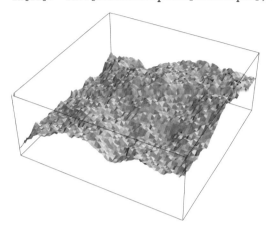

Here are the same data rendered as a contour graphic, showing the level curves on a map of the previous surface.

`In[13]:= Show[ ContourGraphics[landscape0] ];`

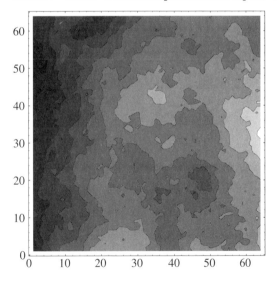

In preparation for a photorealistic rendering of this surface, we increase its resolution and write the height values to a file.

We refine the surface to 128×128 points.

`In[14]:= landscape1 = fBmRA[ landscape0, 128, 0.75 ];`

The appearance of `landscape1` is the same as is the appearance of `landscape0`, but more details are visible.

In[15]:= Show[SurfaceGraphics[landscape1], Mesh -> False];

We write out the data in the form of a pgm file. Such a file is a compact way to store height data. Each point is represented by a single byte, allowing for 256 different height values. The `WritePGM` function, too, is defined in fBm.m.

In[16]:= WritePGM["/tmp/landscape1.pgm", landscape1,
                   "fBmRA[{{0,0}, {0,0}}, 128, 0.75]"]

Out[16]= /tmp/landscape1.pgm

We want to render the surface with ray tracing (see Chapter 11). The program POVRAY requires the height data in the form of a GIF file. We can convert our pgm file into GIF format easily:

```
ppmtogif -sort landscape1.pgm > landscape1.gif
```

This conversion function is part of the PBM package, which provides for many conversions between graphic file formats. The `-sort` option sorts the color map. POVRAY does not look at the colors (or gray levels) in the GIF file, but only at the color indices. Sorting a color map of gray levels makes sure that height corresponds to gray level. A rendering of our data with simple textures and the customary water level is shown in Plate 3-a (left). Most surfaces in artificial landscapes are produced with such fBm methods.

8.2.2 Designer Landscapes

One advantage of the random-addition method is that the initial data can be specified exactly. In this way, we can study the effect of various parameters more easily.

We define an initial grid consisting of four points.

In[17]:= h = 0.7;\
         x0 = {{0, 0}, {0, 0.5}};

8.2 Random Additions

Here we perform five random-addition steps and save the intermediate results.

```
In[18]:= NestList[fBmRAStep[#, h]&, SurfaceGraphics[x0], 5];
```

These six surface graphics show the evolution of randomness.

```
In[19]:= Show[ GraphicsArray[ Partition[%, 3] ] ];
```

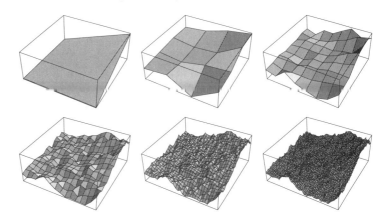

Here we generate data with different values of the lacunarity r, in the form of contour graphics. All pictures use the same initial data.

```
In[20]:= Table[ ContourGraphics[fBmRA[x0, 64, h, r],
            {Contours -> 15, FrameTicks -> None,
             PlotLabel -> StringForm["r = `1`", r]}],
         {r, 0.1, 0.9, 0.8/5} ];
```

The effect of the lacunarity is not easy to describe. With smaller values of r, more points are added at each step and, consequently, fewer steps are required.

```
In[21]:= Show[ GraphicsArray[ Partition[%, 3] ] ];
```

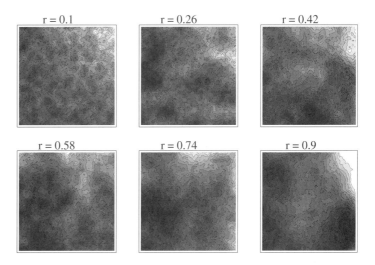

The ability to specify the initial data allows us to produce any desired type of landscape, such as mountains, valleys, or, as shown here, a crater island.

The initial data is a low-resolution plot of a rotated Bessel function.

```
In[22]:= Plot3D[ BesselJ[4, Sqrt[x^2 + y^2]],
                 {x, -7, 7}, {y, -7, 7}, PlotPoints -> 8 ];
```

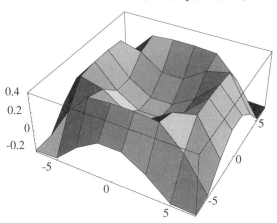

We turn it into a random landscape with 64 × 64 points.

```
In[23]:= fBmRA[%, 64, 0.8];
```

This density graphic shows the resulting height levels. The crater is clearly visible.

```
In[24]:= Show[ DensityGraphics[%], Mesh -> False ];
```

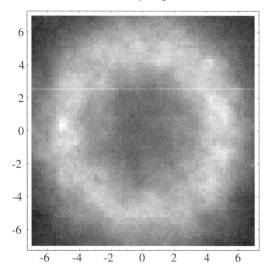

To render it with POVRAY we write it out in pgm form and then convert the data into a GIF file in the same way that we did in Section 8.2.1. The last argument of WritePGM is a string that is written to the pgm file (as documentation).

```
In[25]:= WritePGM[ "/tmp/crater.pgm", %,
                   "BesselJ[4, r], h=0.8" ]
Out[25]= /tmp/crater.pgm
```

8.3 Fourier Synthesis

The result is shown on the right side in Plate 3-a. More sophisticated textures, lighting, and waves give it a more realistic appearance.

8.2.3 Density Functions and Clouds

Here is an example in three dimensions. A tensor of rank 3 can be interpreted as a density function in space.

We generate an $12 \times 12 \times 12$ tensor.

```
In[26]:= grid = fBmRA[Table[0, {2}, {2}, {2}], 12, 0.7];
```

Here we pick all points where the density is above 0.2.

```
In[27]:= Position[grid, x_ /; x > 0.2] // Short
Out[27]//Short=
   {{1, 1, 7}, {1, 1, 8}, {1, 1, 9}, {1, 6, 12}, {1, 9, 4},
    <<86>>, {12, 10, 2}, {12, 11, 1}}
```

We can visualize these points with cubes. Here is a function that generates such lists of cubes for a given threshold value t.

```
In[28]:= cuboids[grid_, t_, opts___] :=
         Graphics3D[Cuboid /@
                    Position[grid, x_ /; x > t, {-1}],
                   {PlotRange -> ({0.99, # + 1.01}& /@
                                  Dimensions[grid]),
                    opts} ]
```

We generate graphics for various values of t.

```
In[29]:= With[{m = Max[grid]},
         Table[cuboids[grid, t,
               PlotLabel -> StringForm["t = `1`", t]],
               {t, 0.8m, 0.2m, -0.6m/5}] ];
```

Here they are. If the resolution were high enough, such graphics could generate realistic clouds.

```
In[30]:= Show[ GraphicsArray[Partition[%, 3]] ];
```

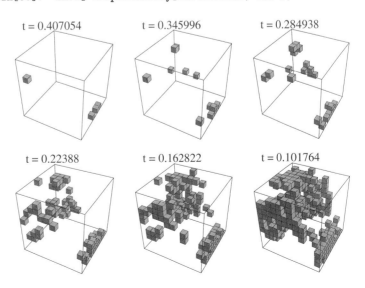

8.3 Fourier Synthesis

The average amplitude v_f of fBm data at frequency f obeys the simple law

$$|v_f| \propto 1/f^{\beta/2} \qquad (8.3\text{--}1)$$

where $\beta = 2h + 1$ is the spectral exponent. One method to generate fBm data is, therefore, to generate random Fourier coefficients distributed according to this formula (with random phase), and then to perform an inverse Fourier transform. To obtain a finite sample, we perform a discrete Fourier transform.

8.3.1 One-Dimensional Synthesis

In one dimension, the Fourier coefficients v_m, for $m = 0, 1, \ldots, n-1$, should have a mean amplitude of

$$m^{-\beta/2}. \qquad (8.3\text{--}2)$$

If we want to generate n points we cannot determine all n coefficients independently, because the result of the inverse Fourier transform should be real valued. For even n, we determine v_0, $v_1, \ldots, v_{n/2}$ according to Equation 8.3–2, with the additional requirement that $v_{n/2}$ be real. The remaining coefficients are $v_i = \overline{v_{n-i}}$, for $i = n/2+1, n/2+2, \ldots, n-1$. Listing 8.3–1 shows the code for `fBmFourier[1, n, h]`.

```
fBmFourier[1, n_?EvenQ, h_, maxn_:Automatic] :=
    Module[{ff, n2 = n/2, mbeta2 = -(2h + 1)/2.0, i, nmax = maxn},
        If[ nmax === Automatic, nmax = n ];
        nmax = Round[nmax/2];
        ff = Table[ freq1[i, mbeta2, nmax], {i, 0, n2} ];
        ff[[-1]] = Re[ff[[-1]]];
        ff = Join[ ff, Reverse[Conjugate[Take[ff, {2, -2}]]] ];
        n Chop[InverseFourier[ff]]
    ]
freq1[0, exp_, nmax_] := 0
freq1[i_, exp_, nmax_] /; i > nmax := 0
freq1[i_, exp_, nmax_] := gauss i^exp randPhase
```

Listing 8.3–1: Part of fBm.m: Fourier synthesis in one dimension

The optional parameter *maxn* limits the highest coefficients to be generated. It is described later. The auxiliary function `freq1` generates the coefficients as random complex numbers with the desired amplitude and uniformly distributed phase.

8.3 Fourier Synthesis

We generate graphs of fBm data for various h in the same way that we did at the beginning of Section 8.2.

```
In[31]:= plots =
           Table[ ListPlot[fBmFourier[1, 1024, h],
               PlotJoined -> True, Axes -> None,
               PlotRange -> All, DisplayFunction -> Identity,
               AspectRatio -> 0.3,
               PlotLabel -> StringForm["h = `1`", h]],
             {h, 0.8, 0.2, -0.2} ];
```

Here are the curves. They should look similar to the ones generated by random additions. There is one apparent difference in overall appearance: Because of the way we constructed the curves, they are periodic and could be put next to each other without any visible anomalies between adjacent copies.

```
In[32]:= Show[ GraphicsArray[Partition[plots, 2]],
             GraphicsSpacing -> {0.1, 0.3} ];
```

8.3.2 Two-Dimensional Synthesis

In two dimensions, the Fourier coefficients $v_{k_1 k_2} = v_{\mathbf{k}}$ should have mean amplitude

$$|\mathbf{k}|^{-(\beta+1)/2} = (k_1^2 + k_2^2)^{-(\beta+1)/4} = (k_1^2 + k_2^2)^{-(h+1)/2}. \qquad (8.3\text{--}3)$$

The symmetry conditions on the two-dimensional coefficients are much more complicated than are the conditions in one dimension. Listing 8.3–2 shows the code of fBmFourier[2, n, h].

We generate a 64 × 64 matrix.   `In[33]:= fBmFourier[2, 64, 0.85];`

Here is the corresponding surface. It has fractal dimension 2.15.   `In[34]:= Show[SurfaceGraphics[%], Mesh -> False];`

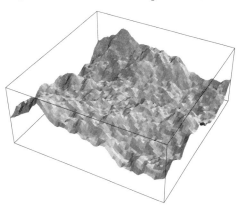

```
fBmFourier[2, n_?EvenQ, h_, maxn_:Automatic] :=
    Module[{line0, half1, linem, half2, full,
            n2 = n/2, mbeta2 = -(h + 1)/2.0, i, j, nmax = maxn},
        If[ nmax === Automatic, nmax = n ];
        nmax = Round[nmax/2];
        line0 = Table[freq2[0, j, mbeta2, nmax], {j, 0, n2}];
        line0[[-1]] = Re[line0[[-1]]];
        line0 = Join[ line0, Conjugate[Reverse[Take[line0, {2, -2}]]] ];
        half1 = Table[ freq2[i, Min[j, n-j], mbeta2, nmax],
                    {i, 1, n2-1}, {j, 0, n-1} ];
        linem = Table[freq2[n2, j, mbeta2, nmax], {j, 0, n2}];
        linem[[1]] = Re[linem[[1]]];
        linem[[-1]] = Re[linem[[-1]]];
        linem = Join[ linem, Conjugate[Reverse[Take[linem, {2, -2}]]] ];
        half2 = RotateLeft /@ half1;
        half2 = Reverse[ Reverse /@ half2 ];
        half2 = Conjugate[half2];

        full = Join[ {line0}, half1, {linem}, half2 ];
        N[Sqrt[n]] Chop[InverseFourier[full]]
    ]
freq2[0, 0, exp_, nmax_] := 0
freq2[i_, j_, exp_, nmax_] /; i > nmax || j > nmax := 0
freq2[i_, j_, exp_, nmax_] := gauss (i^2+j^2)^exp randPhase
```

Listing 8.3–2: Part of fBm.m: Fourier synthesis in two dimensions

This surface has twice the resolution of the previous one, but the same number of nonzero Fourier coefficients. The optional last parameter gives the highest frequency to include. The overall appearance of the surface is similar, therefore, but the added points make it look smoother. The Fourier synthesis method is not incremental as is the random-addition method, so there is no easy way to modify previously generated data.

```
In[35]:= Show[
            SurfaceGraphics[fBmFourier[2, 2 64, 0.85, 64]],
            Mesh -> False ];
```

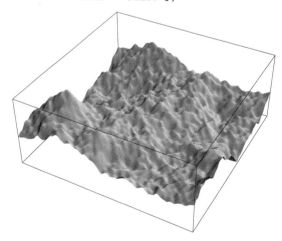

This surface has fractal dimension 2.35.

```
In[36]:= fourierscape = fBmFourier[2, 128, 0.65];
```

8.3 Fourier Synthesis

A version of `fourierscape` with 512 × 512 resolution is shown in Plate 3-b. The higher fractal dimension is typical of rocky mountain terrain. Consequently, we chose a stone texture for the image (adapted from a texture distributed with POVRAY).

```
In[37]:= Show[ SurfaceGraphics[fourierscape],
              Mesh -> False ];
```

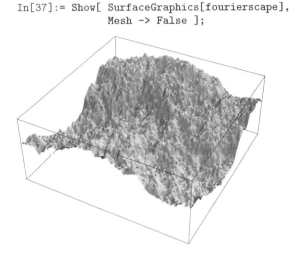

Two-dimensional fBm data can also be used to imitate clouds. The set of points whose values are larger than some threshold has a suitable appearance.

We find the minimal and maximal values in the `fourierscape` data.

```
In[38]:= {min, max} = Through[ {Min, Max}[fourierscape] ]
Out[38]= {-0.986453, 0.772292}
```

This contour graphic has only one contour at the 75 percent level. With some imagination, the white areas can be interpreted as clouds (or as islands; maybe such images are suitable for a Rorschach test?). In fact, the clouds in Plate 3-b were generated in a similar way, using the fBm generator built into POVRAY.

```
In[39]:= Show[ ContourGraphics[fourierscape],
              Contours -> {min + 0.75(max-min)},
              FrameTicks -> None ];
```

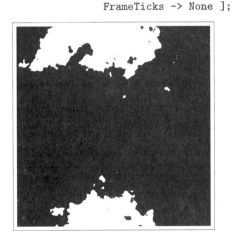

I did not attempt to derive a dimension-independent way to generate Fourier coefficients whose inverse transform is real valued. Note that the function `fBmFourier` still pretends to work for arbitrary dimension e, because it is called in the form `fBmFourier[e, ...]`. But the first parameter is used literally in the two definitions for `fBmFourier[1, ...]` and `fBmFourier[2, ...]`. This setup preserves a consistent user interface.

In Section 12.5.2 we describe another use for fBm data. If we drop the restrictions on the Fourier coefficients, the fBm data will be complex valued, with random phase. We can interpret these complex numbers as color specifications. The command `fBmFourierColor[]` also is part of fBm.m (see Listing 12.5–2).

8.4 Random Faults

The third method can be used to generate periodic fBm data, for example, on a circle or on a sphere. It derives its name from an attempt to model tectonic faults, that is, discontinuities in the crust of the earth. On a circle or on an interval of length T with periodic boundary conditions, the amplitude is written as a sum of random step functions, in the form

$$V(t) = \sum_{i=0}^{k} a_i P(t - t_i), \qquad (8.4\text{--}1)$$

where the a_i are random Gaussian variables, and the t_i are uniformly distributed between 0 and T. The step function is a periodic function with

$$P(t) = \begin{cases} 1 & \text{for } t < T/2 \\ 0 & \text{for } T/2 \leq t < T. \end{cases} \qquad (8.4\text{--}2)$$

The method is easily generalized to multidimensional cases with toroidal symmetry.

8.4.1 Toroidal Symmetry

Our first function `fBmFault[e, n, k]` generates an e-dimensional grid of size n with k random faults. (The grid is initialized to zeroes before faults are added.) Alternatively, we can add faults to an existing grid in the form `fBmFault[`$tensor$`, k]`. `fBmFaultStep[`$tensor$`]` adds a single fault. Let us see how to do this efficiently. In one dimension, we are given a list of length n and must increment a sublist of length $n/2$ (with a random initial position) by a given value. Indices are taken mod n.

Here is a list of length 8 initialized to zeroes.	`In[40]:= n = 8; list = Table[0, {n}];`
We find a random initial position,	`In[41]:= ti = Random[Integer, {1, n}]` `Out[41]= 4`

8.4 Random Faults

and a random increment.	`In[42]:= ai = gauss` `Out[42]= 1.17794`
To avoid complicated index computations, we can simply rotate the list to bring position `ti` to the front. Of course, nothing exciting happens the first time we do this.	`In[43]:= RotateLeft[ list, ti-1 ]` `Out[43]= {0, 0, 0, 0, 0, 0, 0, 0}`
We take the first $n/2$ elements and add the increment, then combine the result with the unmodified remaining elements.	`In[44]:= Join[ Take[%, n/2] + ai, Take[%, -n/2] ]` `Out[44]= {1.17794, 1.17794, 1.17794, 1.17794, 0, 0, 0, 0}`
We put the list back into the original order.	`In[45]:= RotateRight[ %, ti-1 ]` `Out[45]= {0, 0, 0, 1.17794, 1.17794, 1.17794, 1.17794, 0}`

If we work with e-dimensional data, we have to find e random initial positions and increment a fraction of $(1/2)^{(1/e)}$ elements in each dimension (such that exactly half of the total elements are incremented). The code is shown in Listing 8.4–1.

```
fBmFault[mat_List, k_Integer:1] :=
    Nest[ fBmFaultStep, mat, k ]
fBmFault[e_Integer, n_Integer, args__] :=
    fBmFault[ Array[0&, Table[n, {e}]], args ];
fBmFaultStep[mat_List] :=
    With[{dims = Dimensions[mat]},
        hit[ mat, Random[Integer, {1, #}]& /@ dims, 0.5^(1/Length[dims]), gauss ]
    ]
hit[mat_List, {p1_, p___}, frac_, delta_] :=
    Module[{n = Length[mat], mat1, m},
        m = Round[frac n];
        mat1 = RotateLeft[mat, p1-1];
        mat1 = Join[ hit[#, {p}, frac, delta]& /@ Take[mat1, m],
                Take[mat1, -(n - m)] ];
        RotateRight[mat1, p1-1]
    ]
hit[mat_, {}, _, delta_] := mat + delta
```

Listing 8.4–1: fBm.m: Random faults

You may have noticed that the `fBmFault` routines do not take a Hurst exponent as one of their arguments. In fact, the resulting data is always an ordinary random walk with $h = 1/2$. To obtain different values of h, we would have to change the shape of the step function P. Because we have two other good methods to generate such data, I did not bother to implement this complicated case.

Here we show how the random walk looks better and better as more faults are added. We use a list of length 1024 and generate intermediate versions with $1, 4, 4^2, \ldots, 4^5$ faults.

```
In[46]:= FoldList[ fBmFault, fBmFault[1, 1024, 1],
                   4^Range[1, 5] - 4^Range[0, 4] ];
```

We generate list plots.

```
In[47]:= plots =
           MapIndexed[
             ListPlot[#1, PlotJoined -> True, Axes -> None,
                      DisplayFunction -> Identity,
                      PlotRange -> All, AspectRatio -> 0.3,
                      PlotLabel -> StringForm["k = `1`",
                                  4^(First[#2]-1)]]&,
             % ];
```

Here are the curves.

```
In[48]:= Show[ GraphicsArray[Partition[plots, 3]],
               GraphicsSpacing -> {0.1, 0.3} ];
```

k = 1   k = 4   k = 16

k = 64   k = 256   k = 1024

8.4.2 A Functional View of Random Faults

As a final example of fBm code, let us discuss another way to represent fBm data. In all the functions so far, we ended up with a grid of samples of the underlying fractal object. With the fault model we can take another approach: We can generate $V(t)$ in Equation 8.4–1 as a function that can be evaluated for arbitrary t. Let us show how this works for spherical data. In this case, the values of **t** are points on the sphere, given as unit vectors in space, and the $\mathbf{t}_i$ are random points on the sphere, obtained with rand3D. The step function $P(\mathbf{t}, \mathbf{t}_i)$ is equal to 1 in one hemisphere, where $\mathbf{t} \cdot \mathbf{t}_i > 0$, and 0 in the other hemisphere. The function SphericalFaults[k] returns a function corresponding to k random faults. It can be evaluated with any unit vector (point on the sphere) as its argument. The code is shown in Listing 8.4–2.

```
hemi[ti_, delta_] := Function[t, If[ ti.t > 0, delta, 0]]
SphericalFaults[0] = 0&
SphericalFaults[k_Integer?Positive] :=
    With[{sum = Sum[hemi[rand3D, gauss], {k}]},
        Function[t, Evaluate[Through[sum[t], Plus]]]
    ]
```

Listing 8.4–2: Part of fBm.m: Functional spherical faults

8.5 Analysis of fBm Data

The auxiliary function `hemi` generates the single step functions. We form a formal sum of k such functions and use `Through` to distribute the argument `t` to them.

Here is a function corresponding to eight faults.

```
In[49]:= (fault = SphericalFaults[8]) // Short
Out[49]//Short= Function[MathProg`fBm`Private`t$, <<1>>]
```

It can be evaluated for any point on the sphere.

```
In[50]:= fault[ {0,0,1} ]
Out[50]= -2.32164
```

We want to use the tick mark options from this package (see [44]).

```
In[51]:= Needs["Graphics`Graphics`"]
```

Here is a "world map" of these faults in equirectangular projection. The great circles that form the boundaries of the faults turn into sinusoidal curves.

```
In[52]:= DensityPlot[fault[{Sin[theta] Cos[phi],
                           Sin[theta] Sin[phi], Cos[theta]}],
             {phi, 0, 2Pi}, {theta, 0, Pi},
             PlotPoints -> {72, 144}, FrameTicks -> PiScale,
             Mesh -> False, AspectRatio -> Automatic ];
```

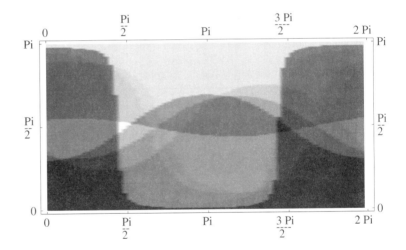

To produce a spherical plot, we load the necessary package.

```
In[53]:= Needs["Graphics`ParametricPlot3D`"]
```

Here, we generate 2000 random faults.

```
In[54]:= fault = SphericalFaults[2000];
```

This command generates a fractal planet. The resolution is rather low and height differences have been exaggerated, so perhaps it looks more like an asteroid. The image is shown on page 163.

```
In[55]:= SphericalPlot3D[ fault[{Sin[theta] Cos[phi],
                                 Sin[theta] Sin[phi],
                                 Cos[theta]}] + 300,
             {theta, 0, Pi}, {phi, 0, 2Pi},
             PlotPoints -> {72, 96}, Ticks -> None ];
```

8.5 Analysis of fBm Data

Finally, let us show how we can analyze the data we generated with the various methods. We want to verify Equation 8.1–2.

We generate test data that should have a Hurst exponent of 0.75.

```
In[1]:= data = fBmRA[1, 2048, 0.75];
```

This function computes the differences of samples dt apart and forms the average over all such differences.

```
In[2]:= delta[data_, dt_] :=
           Plus @@ Abs[Drop[data, -dt] - Drop[data, dt]]/
           (Length[data] - dt)
```

We compute the average values of the differences for dt from 1 to 50.

```
In[3]:= deltas = Table[{dt, delta[data, dt]}, {dt, 50}];
```

Here they are.

```
In[4]:= ListPlot[ deltas, PlotRange -> All ];
```

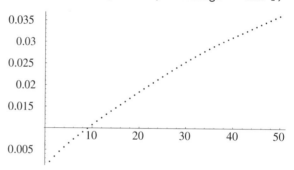

Because h is an exponent, we need nonlinear fitting to estimate it.

```
In[5]:= Needs["Statistics`NonlinearFit`"]
```

We fit the observed average differences to the expected form. You can use NonlinearRegress instead of NonlinearFit if you want to obtain data on the quality of this estimate.

```
In[6]:= NonlinearFit[deltas, a + b dt^h, dt,
           {a, b, h}]

Out[6]= -0.384433 + 0.0633903 dt^0.726831
```

This experiment can be used to gain some confidence in the correctness of our functions. Because they involve randomness, it is not so easy to determine whether the output is valid.

 The CD-ROM contains the package fBm.m, as well as the notebook fBm.nb with the examples from this article.

Chapter Nine

Uniform Polyhedra

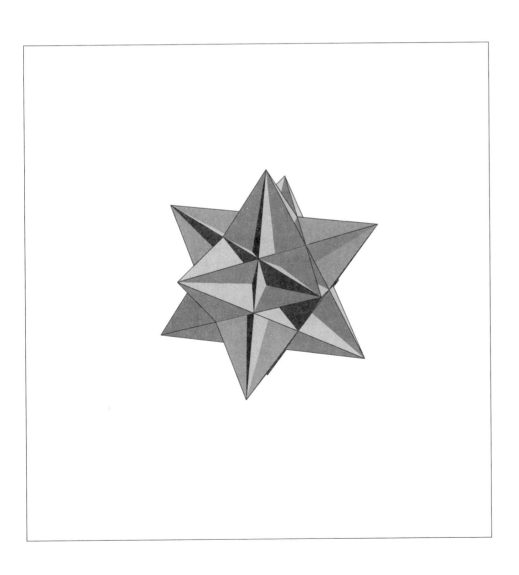

Uniform polyhedra consist of regular faces and congruent vertices. Allowing for nonconvex faces and vertex figures, there are 75 such polyhedra, as well as two infinite families of prisms and antiprisms. A recently discovered uniform way of computing their vertex coordinates is discussed in Section 2. It is the basis for our program to display all of these solids (Sections 3 and 4), among which are many beautiful and stunning shapes. The program requires a flexible way to store properties of objects. We shall also develop some utilities for drawing spherical triangles in Section 5 and for storing properties of objects in a flexible way.

About the illustration overleaf:
The great icosahedron, one of the uniform polyhedra. It is one of four nonconvex regular polyhedra.

The code for this picture is in BookPictures.m.

9.1 Introduction

If all faces of a polyhedron are *regular polygons* (not necessarily all of them the same) and all vertices are *congruent*, we call the polyhedron *uniform*. Regular polyhedra are, of course, also uniform. There are 18 convex uniform polyhedra, namely, the 5 Platonic solids and 13 Archimedian solids. There are also two families of *prisms* and *antiprisms*. One such solid exists for each regular n-gon, for $n = 2, 3, \ldots$ (the case $n = 2$ is degenerate). Figure 9.1–1 shows these 20 solids. We chose $n = 5$ as representative of the prisms and antiprisms.

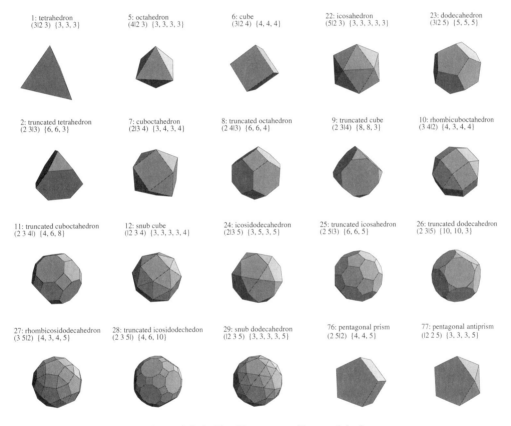

Figure 9.1–1: The 20 convex uniform polyhedra

The first line of the plot labels gives the number of the polyhedron in the standard list in the package PolyhedraExamples.m, and its name. The second line gives its *Wythoff symbol*, described later, and the *vertex configuration*. The vertex configuration describes which polygons are arranged around a vertex and in what order. Because vertices are congruent, this ordering is the same for all vertices. The integer n stands for a regular n-gon. (Later we shall also encounter star polygons, such as the pentagram, which is denoted by $\frac{5}{2}$.) For example, the vertex configuration of the *rhom-*

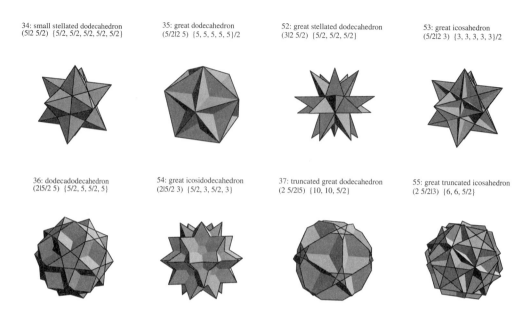

Figure 9.1–2: The four Kepler–Poinsot polyhedra (top row) and related forms (bottom row)

bicuboctahedron (no. 10) is $\{4, 3, 4, 4\}$: three squares and one regular triangle meet at each vertex.

If we drop the condition that the solids be convex, we get many more uniform polyhedra. First, we can allow nonconvex polygons as faces. The nonconvex regular polygons are the *star polygons*. The symbol $\frac{n}{d}$ denotes a star n-gon obtained by connecting every dth vertex of an ordinary n-gon (for example, $\frac{5}{2}$ is the pentagram). Second, we can arrange the faces around a vertex in a nonconvex way. The vertex configuration $\{5, 5, 5, 5, 5\}/2$, for example, says that five pentagons are arranged around the vertex, but the vertex is traversed twice (this polyhedron is the great dodecahedron, no. 35).

With these additional possibilities we get four more regular solids, the *Kepler–Poinsot polyhedra*. They are displayed in Figure 9.1–2 (top row) together with their truncated forms (bottom row). *Truncation* consists of "cutting off" a vertex. For these polyhedra the cut figure is a regular polygon, because the vertex figure is regular. If we cut in such a way that the faces turn into new regular polygons with twice as many edges, we get another uniform polyhedron. If the cuts meet in the middle of the edges, we get yet another solid. The truncated small stellated dodecahedron is an ordinary dodecahedron, and the truncated great stellated dodecahedron turns into an icosahedron. The truncated-in-the-middle forms of the great icosahedron and the small stellated dodecahedron are the same, called the *great icosidodecahedron*, and so are the forms of the great dodecahedron and the great stellated dodecahedron, the *dodecadodecahedron*. The standard package Graphics`Polyhedra` contains an operation Truncate[*graphics3D, ratio*] that performs this operation on any polyhedron. Some of the Archimedian solids can be obtained in this way from the Platonic ones. Here, we do not use these operations.

It turns out that there are exactly 75 uniform polyhedra, other than the infinitely many prisms and antiprisms. The list of all 75 uniform polyhedra was first described by Coxeter and colleagues [12] and later shown to be complete by J. Skilling [55] based on a computer-generated enumeration of all possibilities.

Rather than constructing each of the polyhedra by a special method, we shall find a way to compute the coordinates of any of the uniform polyhedra directly from its Wythoff symbol. First, we briefly describe the constructions for obtaining all uniform polyhedra, then look at the iterative numerical computation of their vertex coordinates, and finally describe the programs in more detail. Much of the material in this chapter has been taken from the paper *Uniform Solution for Uniform Polyhedra* by Zvi Har'El [26]. The programs in UniformPolyhedra.m are a *Mathematica* translation of Har'El's program kaleido.c to compute the polyhedra properties, together with our code for rendering them. The code is reproduced in part throughout this chapter.

9.2 Uniform Construction

All but 1 of the 75 polyhedra can be obtained by a construction involving the *Schwarz triangles*. The construction is described by the Wythoff symbol, which looks like $p|q\,r$, $p\,q|r$, $p\,q\,r|$, or $|p\,q\,r$, where p, q, and r are rational numbers satisfying certain properties.

9.2.1 Schwarz Triangles

Schwarz triangles are spherical triangles (triangles on a sphere) that can be used to cover a sphere exactly by spherical reflections at their edges. Apart from an infinite family of triangles with angles $\pi/2$, $\pi/2$, and π/n, for $n = 2, 3, \ldots$, there is only a finite number of such triangles. The angles must be rational multiples of π, which we write as π/p, π/q, and π/r, where p, q, and r are rational numbers greater than one. It can be shown that finite coverings of the sphere exist only if the numerators of p, q, and r are among $\{2, 3, 4, 5\}$ and that 4 and 5 cannot occur together. The set of rational numbers greater than one with these properties is finite. We denote a Schwarz triangle by $(p\,q\,r)$ (see Figure 9.2–1). The simpler Schwarz triangles allow a single covering of the sphere. They are called *Möbius triangles*. There is an infinite family (2 2 n) of *dihedral* Möbius triangles, plus the three triangles (2 3 3), (2 3 4), and (2 3 5), corresponding to the tetrahedral, octahedral, and icosahedral symmetry groups.

Möbius triangles can be assembled in various ways along their edges to produce larger triangles. Some of these larger triangles can also cover the sphere exactly, but with a multiple covering. For example, the dihedral Möbius triangle (2 2 5), which generates the pentagonal prism, has an area of $\frac{1}{20}$th of the sphere. Twenty copies of it can cover the sphere exactly. It can be doubled to get the triangle (2 2 $\frac{5}{2}$) (recall that the numbers are in the denominator; doubling the angle $\pi/5$ gives

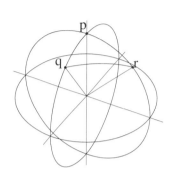

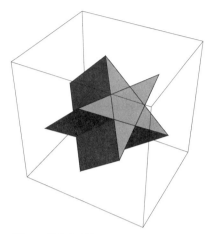

Figure 9.2–1: A Schwarz triangle

Figure 9.2–2: The pentagrammic prism

an angle of $2\pi/5$ or $\frac{\pi}{5/2}$). Twenty of these triangles will cover the sphere *twice*. The associated polyhedron is the *pentagrammic prism* shown in Figure 9.2–2.

9.2.2 The Wythoff Construction

If a point C is chosen in a Schwarz triangle, the images of this point in all the reflections of the triangle that constitute the sphere covering may form the vertices of a uniform polyhedron. All but one of the uniform polyhedra can be obtained in this way. The details can be found in Har'El's paper [26] and in [12]. There are essentially four ways in which a point with this property can be chosen in the triangle $(p\ q\ r)$:

1. C is at a vertex of the triangle, say, at $p = \frac{n}{d}$. This gives a polyhedron with vertex configuration $\{q\ r\ q\ r \ldots q\ r\}$ (with $2n$ terms). Its Wythoff symbol is $p|q\ r$; we use w1[p, q, r] on input. There are 16 of these polyhedra, all regular or *quasiregular* (with just two alternating kinds of faces). These polyhedra are nos. 1, 5, 6, 7, 22, 23, 24, 30, 34, 35, 36, 41, 47, 52, 53, and 54 in Plate 5.

2. C lies on the arc pq and on the bisector of the opposite angle r. This gives a polyhedron with vertex configuration $\{p\ 2r\ q\ 2r\}$. Its symbol is $p\ q|r$; we use w2[p, q, r] on input. There are 33 of these *semiregular* polyhedra. These polyhedra are nos. 2, 3, 4, 8, 9, 10, 13, 14, 15, 17, 19, 25, 26, 27, 31, 33, 37, 38, 42, 43, 44, 48, 49, 51, 55, 58, 61, 62, 65, 66, 67, 70, and 71 in Plate 5. The prisms $2\ n|2$ are also in this class (nos. 76 and 78).

3. C is at the incenter of the triangle. The incenter is the point whose distance from all three edges is the same. This gives a polyhedron with vertex configuration $\{2p\ 2q\ 2r\}$. It is denoted

9.2 Uniform Construction

by $p\ q\ r|$, and w3[p, q, r] on input. There are 14 of these *even-faced* polyhedra. These polyhedra are nos. 11, 16, 18, 20, 21, 28, 39, 45, 50, 56, 59, 63, 68, and 73 in Plate 5.

4. C traces a *snub polyhedron*. This is a polyhedron that is generated by rotations (or an even number of reflections) only. Often it turns out to be chiral, that is, it is different from its mirror image and there is a left and right version of it. The vertex configuration is $\{3\ p\ 3\ q\ 3\ r\}$. The three triangles that connect the p-, q- and r-gons are called *snub triangles*. The snub polyhedra are denoted by $|p\ q\ r$, and w0[p, q, r]. There are 11 of them. These polyhedra are nos. 12, 29, 32, 40, 46, 57, 60, 64, 69, 72, and 74 in Plate 5. The antiprisms and crossed antiprisms $|2\ 2\ n$ are also in this class (nos. 77, 79, and 80).

The auxiliary procedure AnalyzeWythoff that classifies the input is shown in Listing 9.2–1.

```
AnalyzeWythoff[w:(w1|w2|w3|w0)[_, _, _]|w0[_, _, _, _]] :=
    Module[{l = List @@ w, nums, g, k, chi, d},
        If[ Not[ And @@ (l > 1) ], Throw["Wythoff < 1", polyError] ];
        nums = Numerator[l]; k = Max[ nums ];
        If[ Count[l, 2] >= 2,
            (* dihedral *)
                {g, k} = {4k, 2},
            (* tetra, octa, icosa *)
                If[ k > 5, Throw["Wythoff symbol > 5", polyError] ];
                If[ Length[Intersection[nums, {4, 5}]] > 1,
                    Throw["Wythoff symbol contains {4, 5}", polyError] ];
                {g, k} = {24k/(6-k), k}
        ];
        If[ Length[w] === 3,
            chi = g/2 (Plus @@ (1/nums) - 1);
            d   = g/4 (Plus @@ (1/l) - 1);
            If[ d <= 0, Throw["d < 0", polyError] ];
            ,  (* non-Wythoffian: special case, compute later *)
            If[ w =!= w0[3/2, 5/3, 3, 5/2], Throw["unknown Wythoff symbol", polyError] ];
            {chi, d} = {0, 0};
        ];
        {g, k, chi, d}
    ]
```

Listing 9.2–1: Part of UniformPolyhedra.m: Analyzing the Wythoff symbol

Not all polyhedra so constructed are different. There are also cases where some faces degenerate into digons (polygons with just two edges). They can be discarded. For example, the snub cube (no. 12) has the Wythoff symbol |2 3 4. Accordingly, its vertex figure would be $\{3\ 2\ 3\ 3\ 3\ 4\}$. The digon is discarded and we are left with $\{3\ 3\ 3\ 3\ 4\}$, that is, four triangles and one square. Figure 9.2–3 shows the right and left versions of it. Sometimes the remaining polygons no longer form a proper polyhedron. For example, (2|2 5) consists of two pentagons with digons in between. After removing the digons we are left with two coinciding pentagons. (5|2 2) would consist of digons only, that is, a number of lines.

Figure 9.2–3: The two forms of the snub cube

There are some special cases in which a polyhedron constructed by methods 3 or 4 above possesses coinciding faces. In these cases, a modification of the degenerate polyhedron yields another ordinary polyhedron. One such case is $p\ q\ r\,|$, where one of the three numbers, say, r, has an even denominator. This leads to double $2r$-gons which may be discarded. The remaining vertex figure is $\{2p\ 2q\ (2p)'\ (2q)'\}$, where p' denotes a *retrograde* face. Such a face is turned inside out. If you follow the faces around a vertex, you have to back up at a retrograde face. The symbol for a retrograde p-gon is $p' = \frac{p}{p-1}$. For example, a retrograde square is denoted $\frac{4}{3}$ and a retrograde pentagram is $\frac{5}{3}$. Furthermore, these polyhedra are *onesided*, much like a Möbius strip. The vertex figure has a characteristic butterfly-like shape. For example, $2\ 4\ \frac{3}{2}|$ gives a polyhedron named the *small rhombihexahedron*, with vertex configuration $\{8\ 4\ \frac{8}{7}\ \frac{4}{3}\}$.

This command computes a polyhedron from its Wythoff symbol.

```
In[1]:= MakeUniform[w3[2, 4, 3/2]]
Out[1]= MathProg`UniformPolyhedra`Private`uniform$2
```

Here is the vertex configuration of $2\ 4\ \frac{3}{2}|$.

```
In[2]:= VertexConfiguration[%]
Out[2]= {8, 4, 8/7, 4/3}
```

We extended `Graphics3D` to deal with polyhedra and compute their graphic representation. Here is a picture of the small rhombihexahedron.

```
In[3]:= Show[Graphics3D[%%]];
```

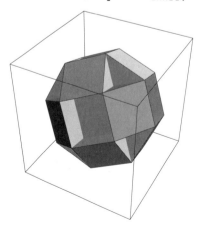

9.2 Uniform Construction

Another special case occurs for the Wythoff symbol $(p\ p'|r)$, where p' is defined as above. Such a polyhedron possesses *hemispherical* faces, that is, faces that pass through the center of the solid. Simple examples are $(\frac{3}{2}\ 3|3)$, the *octahemioctahedron* $\{6\ \frac{3}{2}\ 6\ 3\}$, and $(\frac{3}{2}\ 3|2)$, the *tetrahemihexahedron* $\{4\ \frac{3}{2}\ 4\ 3\}$, shown in Figure 9.2–4. The latter polyhedron is the only one with an odd number of faces (seven), and is also referred to as the *one-sided heptahedron*.

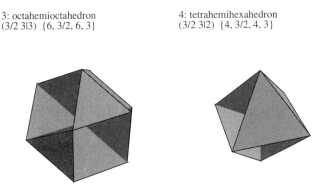

Figure 9.2–4: Two hemispherical polyhedra

There are three polyhedra with star-shaped vertices. Their vertex figures are pentagrams. For example, no. 35, the great dodecahedron, has five pentagons surrounding each vertex twice (see Figure 9.1–2). It is denoted by $\{5\ 5\ 5\ 5\ 5\}/2$. The other two cases are nos. 47 and 53.

9.2.3 The Fundamental Equation

The vertex configuration allows us to compute the spherical angles between the vertices that surround a given vertex. From them we can in turn compute all metric properties of the polyhedron necessary for rendering a graphical image. Although exact formulae exist for all of these polyhedra, they have not been derived in a uniform way and are not necessary for computing vertex coordinates. A numerical approximation serves well for this purpose. In *Mathematica* we can first compute the coordinates to a higher precision and then round them to machine precision for graphic rendering. This is necessary to avoid excessive subdivisions of only approximately planar polygons.

Let the vertex configuration be $(n_1\ n_2\ \ldots\ n_m)$, where the ith face is a regular n_i-gon. There are therefore m faces arranged around any vertex C. The vertices of a uniform polyhedron lie on a sphere and we can project the edges and centers of the faces onto this sphere. Denote by A_i the projection of the center of face i and by B_i the projection of the midpoint of the edge between face i and $i+1$. Figure 9.2–5 illustrates this notation for the cuboctahedron. The set of all the spherical triangles so constructed around C is called the *fundamental system*. It consists of m congruent pairs of triangles $A_iB_{i-1}C$ and A_iB_iC. The angle at B_i is a right angle, the others are denoted by α_i and γ_i (at C).

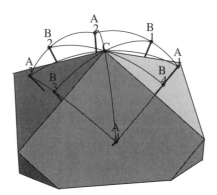

Figure 9.2–5: The fundamental triangles

Because the face A_i is a regular n_i-gon, we get

$$n_i \alpha_i = \pi, \qquad i = 1, 2, \ldots, m. \qquad (9.2\text{–}1)$$

The α_i are, therefore, easy to find. The angles around C must add up to 2π, so

$$\sum_{i=1}^{m} \gamma_i = \pi. \qquad (9.2\text{–}2)$$

All edges have the same length, so the edges a_i (opposite A_i) of the fundamental triangles are all equal in length (the length of spherical edges is given as the angle subtended as viewed from the origin). We call this value a. From spherical trigonometry we get

$$\cos a = \frac{\cos \alpha_i}{\sin \gamma_i}, \qquad i = 1, 2, \ldots, m. \qquad (9.2\text{–}3)$$

Equations 9.2–2 and 9.2–3 together are a system of $m + 1$ equations, called the *fundamental equations*, for the $m + 1$ unknowns γ_i and $\cos a$. We solve it in the procedure `SolveFundamental` with an iterative method developed by Har'El. For reasons of numerical stability, the n_i should be ordered so that the largest one comes first and identical face types should be combined. The code for solving the fundamental equation is shown in Listing 9.2–2.

All other metrical properties such as the circumradius (the distance of vertices from the center), midradius (the distance of edges from the center), inradii (the distances of the faces from the center, different for each type of face), length of edges, dihedral angles, and so on can easily be found from the solutions of the fundamental equations.

9.2.4 Enumeration of Vertices

The next major task is the computation of the *adjacency matrix m*, which lists for each vertex the vertices that surround it, in the proper order. Element m_{ij} of this matrix is the index of the jth

9.2 Uniform Construction

```
SolveFundamental[mi_List, ni_List, d_Integer?NonNegative, prec_] :=
    Module[{alphai, num, pi, cosa, ca1, calphai, delta, g1n, gammai},
        num[x_] := N[x, prec+12]; (* numerical approx *)
        pi = num[Pi];
        (* alpha *)
        alphai = pi/ni; calphai = Cos[alphai];
        ca1 = calphai[[1]];
        calphai = Drop[ calphai, 1 ];

        (* initial values *)
        gammai = num[ pi/2 - alphai ];
        delta = pi d - Plus @@ (mi gammai);

        While[ Abs[delta] > 10^-prec,    (* iterative solution *)
            g1n = gammai[[1]] + delta Tan[gammai[[1]]]/Plus@@(mi Tan[gammai]);
            If[ g1n < 0 || g1n > pi, Throw["gamma out of range", polyError]];
            cosa = ca1/Sin[g1n];
            gammai = Prepend[ ArcSin[ calphai/cosa ], g1n ];
            delta = pi d - Plus @@ (mi gammai);
        ];
        N[{gammai, ca1/Sin[gammai[[1]]]}, prec]
    ]
```

Listing 9.2–2: Part of UniformPolyhedra.m: Solving the fundamental equation

vertex around v_i. We start with the first vertex v_1, chosen on the positive z axis (we set its distance from the origin such that the midradius of the polyhedron will be 1). Because the angle of the edges is $2a$, we can select the second vertex v_2 by a rotation of the first one about this angle. v_2 is the first vertex connected to v_1, so we can set $m_{11} = 2$. We can also set $m_{21} = 1$, because the relation is symmetric. Rotating v_2 by $2\gamma_1$ around v_1 gives the next vertex v_3, and we set $m_{12} = 3$ and $m_{31} = 1$. Then we rotate v_3 by $2\gamma_2$ around v_1, and so on until we rotate v_{m+1} by $2\gamma_m$ to obtain v_2, thus completing the enumeration of vertices around v_1.

We now turn to each vertex v_i in sequence ($i = 2, 3, \ldots$). We know one vertex adjacent to v_i, because (by induction) element m_{i1} has been filled in. Each time we generate a new vertex by a rotation we have to search the list of all known vertices to see whether the new vertex is indeed new or has been encountered before. This search can be done with a simple equality comparison, because we do the computation with higher precision and *Mathematica*'s model of arithmetic guarantees accuracy of all digits within the precision of the numbers involved. The loop stops as soon as i reaches the (known) total number of vertices.

The enumeration is complicated a bit by the fact that, except for snub polyhedra, adjacent vertices have opposite directions because they are related by a reflection of the corresponding Schwarz triangles. Also, we need to keep track of the kind of face the vertices belong to, that is, the element of the vertex configuration ($n_1\ n_2\ldots n_m$) where the enumeration starts. These computations are done in the auxiliary procedure vertices (see Listing 9.2–3).

The adjacency matrix allows the computation of the list of faces needed for generating the polygon data for a graphic rendering.

```
(* adj[[i, j]]: j-th vertex adjacent to vertex i, counterclockwise *)
vertices[poly_] :=
   Module[{v1, v2, vlist, adj, frot, rev, nv, nvi, i, one, start, limit, k, lastk},
      With[{ca = cosa[poly], m = m[poly], v = v[poly], sn = snub[poly],
            gammas = Table[2*gamma[poly][[ rot[poly][[k]] ]], {k, m[poly]}] },
         adj   = Table[0, {v}, {m}];
         frot  = Table[0, {v}];
         vlist = Table[0, {v}];

         v1 = {0, 0, 1}/ca; (* first vertex *)
         frot[[1]] = 1; rev[1] = False; adj[[1, 1]] = 2;
         v2 = {2 ca Sqrt[1 - ca^2], 0, 2 ca^2 - 1}/ca; (* second vertex *)
         If[ snubQ[poly],
                frot[[2]] = If[ sn[[m]], 1, m ]; rev[2] = False; adj[[2, 1]] = 1;
            , (* else reflexible *)
                frot[[2]] = 1; rev[2] = True; adj[[2, m]] = 1;
         ];
         vlist[[1]] = v1; vlist[[2]] = v2; nv = 2;
         i = 1;
         While[ i <= nv, (* more work to do *)
            If[ rev[i],  one = -1; start = m-1; limit = 1,
                         one =  1; start = 2;   limit = m ];
            k = frot[[i]];   (* rotation to use first *)
            v1 = vlist[[i]]; (* the center of rotation *)
            v1 /= Sqrt[norm2[v1]]; (* unit vector *)
            v2 = vlist[[ adj[[i, start - one]] ]];
            Do[ (* rotate previous adjacent vertex *)
                v2 = rotate[ v2, v1, one*gammas[[k]] ];
                (* equality is good enough for comparison *)
                (* searching backwards is faster *)
                For[ nvi = nv, nvi > 0 && v2 != vlist[[nvi]], nvi--, ];
                If[ nvi == 0, nvi = nv + 1 ]; (* new *)
                adj[[i, j]] = nvi;
                lastk = k; k = next[k, m];
                If[ nvi > nv, (* new vertex *)
                    If[ nvi > v, Throw["too many vertices", polyError]];
                    vlist[[nvi]] = v2;
                    If[ snubQ[poly],
                           frot[[nvi]] = If[ !sn[[lastk]], lastk,
                                         If[ !sn[[k]], next[k, m], k ] ];
                           rev[nvi] = False; adj[[nvi, 1]] = i;
                       ,
                           frot[[nvi]] = k;
                           If[ one > 0, rev[nvi] = True; adj[[nvi, m]] = i,
                                        rev[nvi] = False; adj[[nvi, 1]] = i ];
                    ];
                    nv = nvi;
                ];
              ,{j, start, limit, one}];
            i++
```

```
        ];
        If[ nv != v, Throw["not all vertices found", polyError]];
        adjacent[poly] ^= adj;
        vcoord[poly] ^= vlist;
        firstrot[poly] ^= frot;
        Clear[rev];
    ]]
```

Listing 9.2-3: Part of UniformPolyhedra.m: Vertices and adjacency matrix

9.3 Data Structures

In the course of computing the metric properties of the polyhedra we will accumulate many facts about these polyhedra. An ordinary data type is not flexible enough for our purpose, because we generate new data at many points in a lengthy piece of code. One easy way to handle the data is to generate a symbol (inside `Module`) and then attach properties as upvalues:

$$property[symbol] \wedge= value\ .$$

This method can be viewed as setting up a small database of polyhedral properties. The symbol represents the polyhedron and can be given as an argument to functions such as those that render it.

Most of the computations happen inside the auxiliary procedure `makeUniform[`*symbol*, *w*`]`. It is called inside `Catch[..., polyError]` in the exported procedure `MakeUniform[`*w*`]`, where *w* is the Wythoff symbol of the polyhedron to be constructed. This setup allows arbitrary parts of the computation to simply throw an error condition (in the form `Throw[`*message*, `polyError]`) if something goes wrong. We can then test the return value and either return the generated polyhedron or print the error message. The code in `MakeUniform[]` generates also the symbol to be used for storing the properties of the polyhedron. It is created inside `Module[]` and is, therefore, guaranteed to be unique.

```
MakeUniform[w:(w1|w2|w3|w0)[_, _, _]|w0[_, _, _, _]] :=
    Module[{uniform, res},
        res = Catch[makeUniform[uniform, w], polyError];
        If[ res =!= uniform,
            Message[UniformPolyhedra::err, w, res]; $Failed,
            uniform
        ]
    ]
UniformPolyhedra::err = "`1`: `2`"
```

Note the pattern for the argument. It matches expressions with a head of w1, w2, w3, or w0, and the correct number of arguments. A finer analysis of the Wythoff symbol happens inside the auxiliary procedure AnalyzeWythoff (see Section 9.2.2). It generates an error if the argument is not a legal Wythoff symbol, for example (2|4 5), which contains both 4 and 5.

The computation inside makeUniform performs the following steps. The auxiliary procedures that contain the respective code are given in parentheses.

- Check the Wythoff symbol and classify the polyhedron according to symmetry type (dihedral, tetrahedral, octahedral, or icosahedral). This step is performed in AnalyzeWythoff[] (Listing 9.2–1).

- For each of the cases w1, w2, w3, and w0, generate the vertex configurations and treat special cases (makeUniform[]).

- Sort the face types according to descending size and remove trivial faces (digons). This step is necessary to ensure a unique solution of the fundamental equation (sortAndMerge[]).

- Solve the fundamental equations numerically to find the angles between the vertices (SolveFundamental[], Listing 9.2–2).

- Postprocess some special cases. This step involves changing vertex configurations and removing double faces (makeUniform[]).

- Determine the total number of faces, vertices, and edges (count[]).

- Generate the vertex coordinates and the adjacency matrix (vertices[], Listing 9.2–3).

- Compute the list of vertices around each face and the list of faces around each vertex (incidence[]).

The complete code of UniformPolyhedra.m is given on the CD-ROM.

Here is an example. We compute (2|3 4), the cuboctahedron, and show a few of the properties.

```
In[4]:= co = MakeUniform[w1[2, 3, 4]]
Out[4]= MathProg`UniformPolyhedra`Private`uniform$10
```

Here are its symmetry group and the number of edges that meet at each vertex.

```
In[5]:= {SymmetryGroup[co], ValenceOfVertices[co]}
Out[5]= {octahedral, 4}
```

This is the list of vertex numbers for each face. Each sublist has length 3 or 4, because the polyhedron consists of triangles and squares.

```
In[6]:= FaceList[ co ]
Out[6]= {{1, 2, 3}, {1, 3, 8, 4}, {1, 4, 5}, {1, 5, 7, 2},
         {2, 7, 6}, {2, 6, 9, 3}, {3, 9, 8}, {4, 8, 10},
         {4, 10, 11, 5}, {5, 11, 7}, {6, 7, 11, 12}, {6, 12, 9},
         {8, 9, 12, 10}, {10, 12, 11}}
```

9.3 Data Structures

Name	Description
Wythoff	the Wythoff symbol
OrderOfGroup	order of the symmetry group
SymmetryGroup	the name of the symmetry group
Characteristic	the Euler characteristic of the polyhedron
Density	the density (number of times sphere is covered)
NumberOfFaceTypes	number of different types of faces
NumberOfFacesPerType	number each type occurs in vertex configuration
TypeOfFaces	the type of the faces
ValenceOfVertices	number of faces around a vertex
RawConfiguration	the vertex configuration (as index into TypeOfFaces)
VertexConfiguration	the vertex configuration (standard form)
HemisphericalQ	true, if polyhedron includes hemispherical faces
OnesidedQ	true, if nonorientable
SnubQ	true, if a snub polyhedron
CosA	cosine of angle subtended by the half-edges
Gammas	the angles γ_i
NumberOfVertices	total number of vertices
NumberOfEdges	total number of edges
NumberOfFaces	total number of faces
TotalNumberOfFacesPerType	total number of faces of each type
VertexCoordinates	coordinates of all vertices
FaceList	list of vertices of each face
VertexList	list of faces around each vertex
FaceTypes	index into TypeOfFaces for each face
Adjacency	the adjacency matrix
PolyhedronQ	true, if the argument is a valid polyhedron

Table 9.3–1: Properties of polyhedra

Here is the adjacency matrix. The first row, for example, shows that vertex 1 is connected to vertices 2, 3, 4, and 5.

```
In[7]:= Adjacency[ co ]
Out[7]= {{2, 3, 4, 5}, {7, 6, 3, 1}, {2, 9, 8, 1},
        {8, 10, 5, 1}, {4, 11, 7, 1}, {2, 7, 12, 9},
        {2, 5, 11, 6}, {3, 9, 10, 4}, {3, 6, 12, 8},
        {4, 8, 12, 11}, {5, 10, 12, 7}, {11, 10, 9, 6}}
```

Table 9.3–1 lists all functions defined for polyhedra. The predicate PolyhedronQ can be used in a pattern like *proc*[*poly*_Symbol?PolyhedronQ] :=... to test whether the argument of a function is a polyhedron.

One small programming problem, encountered in the code to sort the face types, is worth mentioning. Given a list of numbers, which permutation sorts it into decreasing order, and what is the inverse permutation? (See also Section 1.3.2.)

Here is a sample list of five real numbers.
```
In[1]:= l = {1.2, 5.4, -2.1, 33, 18.5};
```

We attach the index of each number in a list of pairs.
```
In[2]:= Transpose[{l, Range[Length[l]]}]
Out[2]= {{1.2, 1}, {5.4, 2}, {-2.1, 3}, {33, 4}, {18.5, 5}}
```

Now we can sort it.
```
In[3]:= Reverse[ Sort[ % ] ]
Out[3]= {{33, 4}, {18.5, 5}, {5.4, 2}, {1.2, 1}, {-2.1, 3}}
```

We keep only the indices. They describe the original position of the numbers, or the necessary permutation.
```
In[4]:= tr = Transpose[%][[2]]
Out[4]= {4, 5, 2, 1, 3}
```

The inverse permutation is found by asking for the position of the integers from 1 to the length of the list in turn.
```
In[5]:= itr = Position[tr, #][[1,1]]& /@ Range[Length[tr]]
Out[5]= {4, 3, 5, 1, 2}
```

9.4 Rendering

We overloaded Graphics3D[*polyhedron*] to convert a polyhedron into a list of polygons. The code is shown in Listing 9.4–1. Once we know which vertices constitute each face, it is a simple matter to generate the necessary Polygon commands to render the polyhedron. The list of faces is in the same format as in the standard package Graphics'Polyhedra'. Each face is a list of the numbers of its vertices. The face $\{i_1, i_2, \ldots, i_k\}$ can be used as an argument in Part[*vertices, face*] or *vertices*[[*face*]] to give the list of vertex coordinates to be wrapped in Polygon[*vertexlist*].

To show how this works, we use symbolic names v1, v2, ... for the vertices, instead of coordinates.
```
In[1]:= vertices = {v1, v2, v3, v4};
```

If a face consists of vertices 1, 2, and 4, we can simply extract the corresponding elements to get the list of its vertex coordinates.
```
In[2]:= vertices[[{1, 2, 4}]]
Out[2]= {v1, v2, v4}
```

As a simple example, the tetrahedron has these four faces.
```
In[3]:= faces = FaceList[ MakeUniform[w1[3,2,3]] ]
Out[3]= {{1, 2, 3}, {1, 3, 4}, {1, 4, 2}, {2, 4, 3}}
```

We can use Map to get the list of all polygons.
```
In[4]:= Function[ face, vertices[[face]] ] /@ faces
Out[4]= {{v1, v2, v3}, {v1, v3, v4}, {v1, v4, v2},
         {v2, v4, v3}}
```

9.4 Rendering

Finally, we wrap Polygon around each one.

```
In[5]:= Polygon /@ %
Out[5]= {Polygon[{v1, v2, v3}], Polygon[{v1, v3, v4}],
         Polygon[{v1, v4, v2}], Polygon[{v2, v4, v3}]}
```

If a polygon is not convex, this simple construction does not work because *Mathematica* does not render nonconvex polygons correctly (the problem is not easy to solve in general). Therefore, we wrote a procedure Convexify[] (Listing 9.4–1) to decompose such a polygon into convex parts. We do it in a "natural" way by cutting along the interior parts of its edges. It turns out that the chosen cutting lines are most often lines of intersection of different faces anyway, which helps in rendering the solid. We implemented this method only for star polygons of the form $n/2$ and $n/3$, as well as their retrograde siblings $n/(n-2)$ and $n/(n-3)$. Note that a retrograde face $n/(n-1)$

```
Graphics3D[poly_Symbol?PolyhedronQ, opts___?OptionQ] :=
    With[{fcoord = vcoord[poly][[#]]& /@ faces[poly], ftypes = ni[poly][[ftype[poly]]]},
        Graphics3D[ Apply[Convexify, Transpose[{fcoord, ftypes}], {1}], opts ]
    ]
Convexify[vlist_List, p_] /; Denominator[compl[p]] < Denominator[p] :=
    Convexify[Reverse[vlist], compl[p]]
Convexify[vlist_List, p_] /; Denominator[p] == 2 :=
    Module[{m = Numerator[p], n = Denominator[p], k, v2list, tria, cent},
        k = 1 + Mod[n Ceiling[p], m] Floor[p];
        v2list = {vlist, RotateLeft[vlist, 1],
                  RotateLeft[vlist, -k], RotateLeft[vlist, k]};
        v2list = Apply[ inter, Transpose[v2list], {1} ];
        tria = Transpose[ {vlist, v2list, RotateLeft[v2list, -k]} ];
        cent = {v2list[[ Mod[k Range[0, m-1], m] + 1 ]]};
        Polygon /@ Join[ tria, cent ]
    ]
Convexify[vlist_List, p_] /; Denominator[p] == 3 :=
    Module[{m = Numerator[p], n = Denominator[p], k, quadra, v2list, tria, v3list, cent},
        k = 1 + Mod[n Ceiling[p], m] Floor[p];
        v2list = {vlist, RotateLeft[vlist, 1],
                  RotateLeft[vlist, -2k], RotateLeft[vlist, k]};
        v2list = Apply[ inter, Transpose[v2list], {1} ];
        v3list = {RotateLeft[vlist, -2k], RotateLeft[vlist, k],
                  RotateLeft[vlist, -k], RotateLeft[vlist, 2k]};
        v3list = Apply[ inter, Transpose[v3list], {1} ];
        quadra = Transpose[ {vlist, v2list, v3list, RotateLeft[v2list, -k]} ];
        tria = Transpose[ {v2list, RotateLeft[v3list, k], v3list} ];
        cent = {v3list[[ Mod[k Range[0, m-1], m] + 1 ]]};
        Polygon /@ Join[ quadra, tria, cent ]
    ]
Convexify[vlist_List, p_] := Polygon[vlist] (* integer and catchall *)
```

Listing 9.4–1: Part of UniformPolyhedra.m: Rendering polyhedra

is convex and needs no special treatment. Figure 9.4–1 shows the decomposition of $\frac{5}{2}$ and $\frac{8}{3}$ as examples.

Figure 9.4–1: Convex decomposition of star polygons

Some of the more complicated of these polyhedra (nos. 69, 72, 74, and 75) cannot be rendered with the current version of *Mathematica*'s built-in POSTSCRIPT generator Display. On workstations with special graphics hardware the command Live[*graphics3D*] can be used to render them with the resident hardware and software.

9.4.1 A Polyhedral Gallery

You will discover many beautiful and incredible shapes among the 75 uniform polyhedra. All 75 polyhedra, as well as five examples of prisms and antiprisms (nos. 76–80) are shown in Plate 5. They were ray traced with POVRAY, using the tools developed in Chapter 11. Each type of face has a distinct color. Hexagons, for example, are yellow.

The package PolyhedraExamples.m contains the standard list of these polyhedra and the functions NumberedPolyhedron[n] to compute the nth polygon and ShowPolyhedron[n] to render it complete with a label showing its name, Wythoff symbol, and vertex configuration (Listing 9.4–2). We have used this command for many of the pictures in this chapter. Table 9.4–1 gives the numbers, Wythoff symbols, and names of all 80 solids as defined in PolyhedraExamples.m.

9.4.2 The Last Polyhedron

If you looked at the number of polyhedra for each of the four constructions you might have noticed that their sum equals 74, not 75. There is one uniform polyhedron that cannot be constructed with Wythoff's method. It is the only one with more than six faces around a vertex (it has 8 of them!). It has vertex configuration $\{4\,\frac{5}{3}\,4\,3\,4\,\frac{5}{2}\,4\,\frac{3}{2}\}$. It is denoted by the pseudo Wythoff symbol $(|\frac{3}{2}\,\frac{5}{3}\,3\,\frac{5}{2})$ (w0[3/2, 5/3, 3, 5/2]) and named the *great dirhombicosidodecahedron*. It can be found by a modification of $(|3\,\frac{5}{2}\,\frac{5}{3})$. This remarkable solid which is no. 75 in our list is shown in a ray-traced image on Plate 8. It consists of 40 triangles (20 of them retrograde), 60 hemispherical snub squares, and 24 pentagrams (12 of them retrograde). It has the largest number of faces (124) and edges (240) of all uniform polyhedra. The squares come in 20 coplanar pairs, that is, two

9.4 Rendering

No.	Symbol	Name
1	$3\|2\ 3$	tetrahedron
2	$2\ 3\|3$	truncated tetrahedron
3	$\frac{3}{2}\ 3\|3$	octahemioctahedron
4	$\frac{3}{2}\ 3\|2$	tetrahemihexahedron
5	$4\|2\ 3$	octahedron
6	$3\|2\ 4$	cube
7	$2\|3\ 4$	cuboctahedron
8	$2\ 4\|3$	truncated octahedron
9	$2\ 3\|4$	truncated cube
10	$3\ 4\|2$	rhombicuboctahedron
11	$2\ 3\ 4\|$	truncated cuboctahedron
12	$\|2\ 3\ 4$	snub cube
13	$\frac{3}{2}\ 4\|4$	small cubicuboctahedron
14	$3\ 4\|\frac{4}{3}$	great cubicuboctahedron
15	$\frac{4}{3}\ 4\|3$	cubohemioctahedron
16	$\frac{4}{3}\ 3\ 4\|$	cubitruncated cuboctahedron
17	$\frac{3}{2}\ 4\|2$	great rhombicuboctahedron
18	$\frac{3}{2}\ 2\ 4\|$	small rhombihexahedron
19	$2\ 3\|\frac{4}{3}$	stellated truncated hexahedron
20	$\frac{4}{3}\ 2\ 3\|$	great truncated cuboctahedron
21	$\frac{4}{3}\ \frac{3}{2}\ 2\|$	great rhombihexahedron
22	$5\|2\ 3$	icosahedron
23	$3\|2\ 5$	dodecahedron
24	$2\|3\ 5$	icosidodecahedron
25	$2\ 5\|3$	truncated icosahedron
26	$2\ 3\|5$	truncated dodecahedron
27	$3\ 5\|2$	rhombicosidodecahedron
28	$2\ 3\ 5\|$	truncated icosidodechedron
29	$\|2\ 3\ 5$	snub dodecahedron
30	$3\|\frac{5}{2}\ 3$	small ditrigonal icosidodecahedron
31	$\frac{5}{2}\ 3\|3$	small icosicosidodecahedron
32	$\|\frac{5}{2}\ 3\ 3$	small snub icosicosidodecahedron
33	$\frac{3}{2}\ 5\|5$	small dodecicosidodecahedron
34	$5\|2\ \frac{5}{2}$	small stellated dodecahedron
35	$\frac{5}{2}\|2\ 5$	great dodecahedron
36	$2\|\frac{5}{2}\ 5$	dodecadodecahedron
37	$2\ \frac{5}{2}\|5$	truncated great dodecahedron
38	$\frac{5}{2}\ 5\|2$	rhombidodecadodecahedron
39	$2\ \frac{5}{2}\ 5\|$	small rhombidodecahedron
40	$\|2\ \frac{5}{2}\ 5$	snub dodecadodecahedron
41	$3\|\frac{5}{3}\ 5$	ditrigonal dodecadodecahedron
42	$3\ 5\|\frac{5}{3}$	great ditrigonal dodecicosidodecahedron
43	$\frac{5}{3}\ 3\|5$	small ditrigonal dodecicosidodecahedron
44	$\frac{5}{3}\ 5\|3$	icosidodecadodecahedron
45	$\frac{5}{3}\ 3\ 5\|$	icositruncated dodecadodecahedron
46	$\|\frac{5}{3}\ 3\ 5$	snub icosidodecadodecahedron
47	$\frac{3}{2}\|3\ 5$	great ditrigonal icosidodecahedron
48	$\frac{3}{2}\ 5\|3$	great icosicosidodecahedron
49	$\frac{3}{2}\ 3\|5$	small icosihemidodecahedron
50	$\frac{3}{2}\ 3\ 5\|$	small dodecicosahedron
51	$\frac{5}{4}\ 5\|5$	small dodecahemidodecahedron
52	$3\|2\ \frac{5}{2}$	great stellated dodecahedron
53	$\frac{5}{2}\|2\ 3$	great icosahedron
54	$2\|\frac{5}{2}\ 3$	great icosidodecahedron
55	$2\ \frac{5}{2}\|3$	great truncated icosahedron
56	$2\ \frac{5}{2}\ 3\|$	rhombicosahedron
57	$\|2\ \frac{5}{2}\ 3$	great snub icosidodecahedron
58	$2\ 5\|\frac{5}{3}$	small stellated truncated dodecahedron
59	$\frac{5}{3}\ 2\ 5\|$	truncated dodecadodecahedron
60	$\|\frac{5}{3}\ 2\ 5$	inverted snub dodecadodecahedron
61	$\frac{5}{2}\ 3\|\frac{5}{3}$	great dodecicosidodecahedron
62	$\frac{5}{3}\ \frac{5}{2}\|3$	small dodecahemicosahedron
63	$\frac{5}{3}\ \frac{5}{2}\ 3\|$	great dodecicosahedron
64	$\|\frac{5}{3}\ \frac{5}{2}\ 3$	great snub dodecicosidodecahedron
65	$\frac{5}{4}\ 5\|3$	great dodecahemicosahedron
66	$2\ 3\|\frac{5}{3}$	great stellated truncated dodecahedron
67	$\frac{5}{3}\ 3\|2$	great rhombicosidodecahedron
68	$\frac{5}{3}\ 2\ 3\|$	great truncated icosidodecahedron
69	$\|\frac{5}{3}\ 2\ 3$	great inverted snub icosidodecahedron
70	$\frac{5}{3}\ \frac{5}{2}\|\frac{5}{3}$	great dodecahemidodecahedron
71	$\frac{3}{2}\ 3\|\frac{5}{3}$	great icosihemidodecahedron
72	$\|\frac{3}{2}\ \frac{3}{2}\ \frac{5}{2}$	small retrosnub icosicosidodecahedron
73	$\frac{3}{2}\ \frac{5}{3}\ 2\|$	great rhombidodecahedron
74	$\|\frac{3}{2}\ \frac{5}{3}\ 2$	great retrosnub icosidodecahedron
75	$\|\frac{3}{2}\ \frac{5}{3}\ 3\ \frac{5}{2}$	great dirhombicosidodecahedron
76	$2\ 5\|2$	pentagonal prism
77	$\|2\ 2\ 5$	pentagonal antiprism
78	$2\ \frac{5}{2}\|2$	pentagrammic prism
79	$\|2\ 2\ \frac{5}{2}$	pentagrammic antiprism
80	$\|2\ 2\ \frac{5}{3}$	pentagrammic crossed antiprism

Table 9.4–1: Names and Wythoff symbols of all uniform polyhedra

squares each lie in the same plane, rotated against each other by 38.17 degrees. Because they pass through the center of the solid they do not have an outside normal direction. A consistent normal direction was arbitrarily chosen to make sure that the coplanar faces have the same normal vector. The same method is also needed for other one-sided polyhedra. The triangles and pentagrams also come in coplanar pairs.

```
BeginPackage["MathProg`PolyhedraExamples`", "MathProg`UniformPolyhedra`"]
NumberedPolyhedron::usage = "NumberedPolyhedron[n] gives polyhedron nr. n
        from the standard list."
ShowPolyhedron::usage = "ShowPolyhedron[n] shows a labeled graphic of polyhedron nr. n."
standard::usage = "standard[[i]] gives name and Wythoff symbol of the ith
        polyhedron from the list."
Begin["`Private`"]
(* standard list of {name, Wythoff symbol} *)
standard = {
  {"tetrahedron", w1[3, 2, 3]},
  {"truncated tetrahedron", w2[2, 3, 3]},
    ⋮
  {"pentagrammic crossed antiprism", w0[2, 2, 5/3]}
};
NumberedPolyhedron[n_Integer] /; 1 <= n <= Length[standard] :=
        MakeUniform[ standard[[n, 2]] ]
(* graphic with plot label *)
ShowPolyhedron[n_, opts___?OptionQ] /; 1 <= n <= Length[standard] :=
    Module[{poly, name, conf, wyt},
        poly = NumberedPolyhedron[n];
        name = standard[[n, 1]];
        conf = VertexConfiguration[poly];
        wyt  = Wythoff[poly];
        Show[ Graphics3D[poly,
                    PlotLabel -> ToString[StringForm["`1`: `2`\n", n, name]] <>
                        ToString[StringForm["(`1`)   `2`", wyt, InputForm /@ conf]]
                ],
                opts ]
    ]
End[]
EndPackage[]
```

Listing 9.4–2: PolyhedraExamples.m (excerpt): Definitions for all uniform polyhedra

9.5 Auxiliary Programs

A number of useful utilities for drawing spherical triangles and great circle arcs is available in the package SphericalTriangles.m, reproduced in Listing 9.5–1. The main problem solved in this package is the generation of the graphic object for an arc of a great circle. Such an arc is given by its two endpoints on the unit sphere. From the two points we derive a parametric equation for the great circle that goes through the points. A generally useful method for generating a Graphics3D object from such an equation is to use ParametricPlot3D with the option DisplayFunction->Identity to suppress rendering. The desired list of lines is then the first element of the resulting graphic object.

The function GreatCircleArc[A, B, α] returns a list of graphics primitives that represent the arc starting at A, running in the direction toward B, and having angle α. If the angle is omitted, it is computed such that the arc ends at B. If we set $\alpha = 2\pi$ we get a full great circle. The code makes use of auxiliary functions unit[$vector$] and Cross[v_1, v_2] to compute unit vectors and cross products, respectively.

It is a now a simple matter to draw spherical triangles with given angles α, β, and γ in a function SphericalTriangle[α, β, γ, $options$...] (see Listing 9.5–1). We choose the first vertex at $(0, 0, 1)$ and the second one in the x–z plane. Then we draw the circle arcs, points at the vertices, and lines from the origin to each vertex. An option PointLabels can be used to specify the labels of the three vertices, and GreatCircles->True causes the full great circles of the edges to be drawn.

We redefine also the standard graphics option Boxed to draw a "box" consisting of the three great circles in the coordinate planes and the option Axes to render the x-, y-, and z-axes through the origin. This style of box and axes is more appropriate for spherical plots. Many of the techniques for processing options described in [40] are used here.

The function SchwarzTriangle[p, q, r, $opts$...] draws Schwarz triangles. Figure 9.2–1 was produced with

    Show[SchwarzTriangle[2, 3, 4, PointLabels -> {p, q, r}]].

The CD-ROM contains the packages UniformPolyhedra.m, PolyhedraExamples.m, and SphericalTriangles.m, as well as the notebook UniformPolyhedra.nb, with the examples from this chapter.

```
BeginPackage["MathProg`SphericalTriangles`"]

SphericalTriangle::usage = "SphericalTriangle[alpha, beta, gamma, opts...]
    draws a spherical triangle with the given angles."
SchwarzTriangle::usage = "SchwarzTriangle[p, q, r, opts...]
    draws the Schwarz triangle (p q r)."
GreatCircleArc::usage = "GreatCircleArc[a, b, (w)] draws a great circle arc
    from a in the direction of b. The default for w lets the arc end at b."
GreatCircle::usage = "GreatCircle[a, b] draws a great circle in the plane
    defined by a and b."
PointLabels::usage = "PointLabels->{la, lb, lc} is an option of
    SphericalTriangle to give labels to its three vertices."
GreatCircles::usage = "GreatCircles -> True/False is an option of SphericalTriangle
    that specifies whether full great circles should be drawn for the sides."

Begin["`Private`"]

Needs["Geometry`Rotations`"]
Needs["Utilities`FilterOptions`"]

pi = N[Pi];
pi2 = pi/2;
X = {1,0,0}; Y = {0,1,0}; Z = {0,0,1}; o = {0,0,0}

thin = Thickness[0.001];
thick = Thickness[0.006];
psize = PointSize[0.01];
pr = 1.2            (* protrusion of axes and default plot range *)
vp = 2*{2.4, 1.3, 2} (* default viewpoint *)

Options[SphericalTriangle] = {
    PointLabels->None,
    GreatCircles->False,
    ViewPoint -> vp,
    PlotRange -> pr{{-1, 1}, {-1, 1}, {-1, 1}}
}

SphericalTriangle[alpha_, beta_, gamma_, opts___] /; Pi < alpha + beta + gamma < 3Pi :=
    Module[{a, b, c, A, B, C, lines, points, triangle, pl},
    (* sides *)
    {a, b, c} = Apply[ N[ArcCos[(Cos[#1] + Cos[#2] Cos[#3]) Csc[#2] Csc[#3]]]&,
                NestList[ RotateLeft, {alpha, beta, gamma}, 2 ], {1} ] /.
                ArcCos[Cos[x_]] :> x;
    (* points *)
    A = {0, 0, 1};
    B = Rotate3D[A, 0, c, pi2];
    C = Rotate3D[A, 0, b, pi2 - alpha];
    lines = {thin, Line[{o, A}], Line[{o, B}], Line[{o, C}]};
    triangle = {thick, GreatCircleArc[A, B],
                GreatCircleArc[B, C], GreatCircleArc[C, A]};
    If[ GreatCircles /. {opts} /. Options[SphericalTriangle],
        triangle = Join[{thin, GreatCircle[A, B],
                    GreatCircle[B, C], GreatCircle[C, A]}, triangle] ];
    points = {psize, Point[A], Point[B], Point[C]};
```

9.5 Auxiliary Programs

```
    pl = PointLabels /. {opts} /. Options[SphericalTriangle];
    If[ Head[pl] === List && Length[pl] == 3,
        points = Join[points, MapThread[Text, {pl, {A, B, C},
                                       {{1,-1.5}, {1.5,-1}, {-1.5, -1}}}]
                ] ];
    frame = {thin};
    If[ Boxed /. {opts} /. Options[Graphics3D],
        frame = Join[frame,
                {GreatCircle[X, Y], GreatCircle[Y, Z], GreatCircle[Z, X]}] ];
    ax = Axes /. {opts} /. Options[Graphics3D];
    If[ Head[ax] =!= List, ax = {ax, ax, ax} ];
    If[ Length[ax] == 3,
        If[ ax[[1]], AppendTo[frame, Line[{-pr X, pr X}]] ];
        If[ ax[[2]], AppendTo[frame, Line[{-pr Y, pr Y}]] ];
        If[ ax[[3]], AppendTo[frame, Line[{-pr Z, pr Z}]] ];
    ];
    Graphics3D[{frame, lines, triangle, points},
            { Boxed->False, Axes->False, FilterOptions[Graphics3D, opts],
                FilterOptions[Graphics3D, Sequence @@ Options[SphericalTriangle]]
            }]
    ]
SchwarzTriangle[p_, q_, r_, opts___] /; 1 < 1/p + 1/q + 1/r < 3 :=
    SphericalTriangle[Pi/p, Pi/q, Pi/r, opts]
GreatCircleArc[A_, B_, alpha_] :=
    Module[{au = unit[A], n1, n2, t},
        n1 = Cross[A, B];
        n2 = unit[Cross[n1, A]];
        ParametricPlot3D[ Evaluate[Cos[t] au + Sin[t] n2], {t, 0, alpha},
                    DisplayFunction->Identity ][[1]]
    ]
GreatCircleArc[A_, B_] := GreatCircleArc[A, B, ArcCos[unit[A].unit[B]]]
GreatCircle[A_, B_] := GreatCircleArc[A, B, 2Pi]
(* norms of vector, unit vectors *)
norm2[v_] := Plus@@(v^2)
norm[v_]  := Sqrt[norm2[v]]
unit[v_]  := v/norm[v]
(* rotation axis must be unit vector! *)
rotate[v_, axis_, a_] :=
    With[{p = axis(axis.v), q = Cross[axis, v]}, p + Cos[a] (v - p) + Sin[a] q ]
End[]
Protect[SphericalTriangle, SchwarzTriangle, GreatCircleArc, GreatCircle]
EndPackage[]
```

Listing 9.5–1: SphericalTriangles.m

Chapter Ten

The Stellated Icosahedra

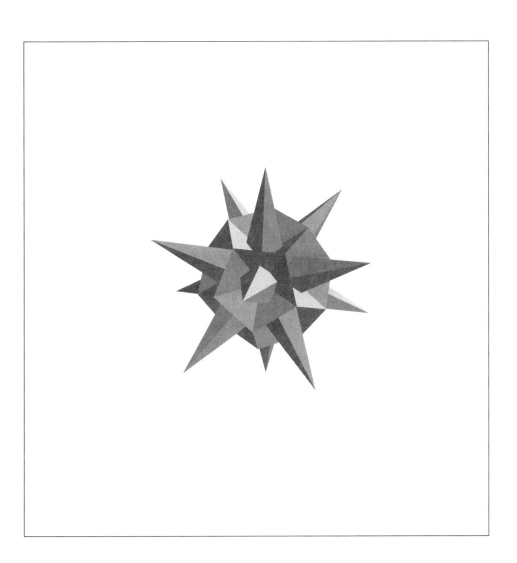

The enumeration of all stellations of the icosahedron was accomplished in 1938. The geometric constructions and combinatorial algorithms used can easily be programmed in *Mathematica*. Its symbolic and graphic capabilities make it well suited to render the solids in a variety of formats.

Section 1 introduces stellations of regular polyhedra and discusses stellations of the octahedron and dodecahedron. Section 2 shows how all stellations of the icosahedron can be generated and represented as symbolic objects in *Mathematica*. In Section 3 we turn to the numerical aspects and generate three-dimensional graphic objects for the stellations. At the end, we discuss some of the interesting stellations.

About the illustration overleaf:

Stellation 23 of the icosahedron. Commands for drawing such pictures are described in this chapter.

The code for this picture is in BookPictures.m.

10.1 Introduction

Stellation refers to the extension the faces of a polyhedron until they intersect with the extensions of other faces, such that the symmetry of the original polyhedron is preserved. The set of all possible edges of the stellated forms can be obtained by finding all intersections of the plane of one face with the planes of all other faces. No stellations are possible for the tetrahedron and the cube. The only stellation of the octahedron is the *stella octangula*, a compound of two dual tetrahedra. The two tetrahedra are related to each other by inversion and can be drawn easily.

The package Graphics`Polyhedra` contains all regular polyhedra; the Graphics`Shapes` package contains the function `AffineShape[]` needed for inverting the tetrahedron.

```
In[1]:= Needs["Graphics`Polyhedra`"]; \
        Needs["Graphics`Shapes`"];
```

The inversion is an affine transformation that multiplies the coordinates of all vertices by -1.

```
In[2]:= tetra = Polyhedron[ Tetrahedron ];\
        dual  = AffineShape[ tetra, {-1, -1, -1} ];
```

Here is the stella octangula. The octahedron that generates this stellation is the intersection of the two tetrahedra.

```
In[3]:= Show[ {tetra, dual} ];
```

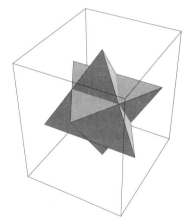

The dodecahedron admits three stellations, discovered by Kepler in 1619 (the *small stellated dodecahedron* and the *great stellated dodecahedron*) and by Poinsot in 1809 (the *great dodecahedron*). They are all regular (but have nonconvex faces or vertices). A *regular* polyhedron has faces that are congruent regular polygons (triangles, squares, and pentagons are the only ones that occur). All vertices show the same arrangement of faces around them. The only nonconvex polygon than can occur as a face of a regular (nonconvex) polyhedron is the pentagram (five-pointed star).

Together with one stellation of the icosahedron—the *great icosahedron*—the stellated dodecahedra form all nonconvex regular polyhedra. The four regular stellated polyhedra are shown here.

See also Chapter 9 for more about uniform polyhedra, to which the regular polyhedra belong.

Polyhedron[*name*] generates a Graphics3D object of the named polyhedron.

```
In[4]:= starpoly = Map[ Polyhedron,
                {{SmallStellatedDodecahedron,
                  GreatStellatedDodecahedron},
                 {GreatDodecahedron,
                  GreatIcosahedron}}, {2}
            ];
```

Color illustrations of these solids can be found on Plate 8. They are nos. 34, 52, 35, and 54, respectively.

```
In[5]:= Show[ GraphicsArray[starpoly, GraphicsSpacing->0] ];
```

10.2 Stellations of the Icosahedron

The only regular polyhedron whose stellations were not completely known in the nineteenth century is the icosahedron. The enumeration of all stellations of the icosahedron was accomplished in 1938 by Coxeter and Du Val. The results were published in the University of Toronto Studies (Mathematical Series). The paper contained isometric drawings by Petrie and planar drawings by Du Val. The fourth author of that paper, Flather, prepared models of all 59 solids. Their whereabouts are explained in the preface of a 1982 Springer-Verlag reprint of the 1938 paper [11]. Much of the material in this chapter is based on this source.

We shall use these option settings throughout the rest of this chapter.

```
In[1]:= SetOptions[GraphicsArray, GraphicsSpacing -> 0];\
        SetOptions[Graphics3D, Boxed -> False];
```

10.2 Stellations of the Icosahedron

All code is part of the package Icosahedra.m, reproduced at the end of this chapter in Listing 10.5–1.

```
In[2]:= Needs["MathProg`Icosahedra`"]
```

10.2.1 Subdivision of the Plane

The subdivision of the plane of an icosahedral face by the intersections of all other planes (except the opposite plane, which is parallel to it) is shown in Figure 10.2–1. This graphic is the value of the variable `lineGraphics` from Icosahedra.m. Assigning the graphic to a variable is a convenient way to access this complicated illustration. Because the outermost points are rather far away, details of the complicated interior intersections are lost. The variable `lineGraphicsInner` is another graphic that shows only the interior points and lines (see Figure 10.2–2).

All points of intersection between the various lines have either threefold rotational symmetry or sixfold dihedral symmetry. (The dihedral group of order six consists of three rotations and three reflections.) The three rotationally symmetric points are labeled a_1, a_2, and a_3, for example. The dihedral points come in two sets of three: left and right. They are labeled bl_1, bl_2, bl_3, and br_1, br_2, br_3. The face of the original icosahedron is the innermost triangle with vertices j_1, j_2, and j_3.

To construct the intersection figure, we started with the equilateral triangle $c_1c_2c_3$. Its edges are then divided in the golden ratio, giving rise to points dl and dr, respectively. All other points can be defined as intersections of lines passing through the points c_i, dl_j, and dr_k. We wrote a command define[*sym*, *coord*] that defines *sym*[1] to be *coord* and *sym*[2] and *sym*[3] to be images of *coord* under successive rotation by 120 degrees. The command inter[*a*, *b*, *c*, *d*]

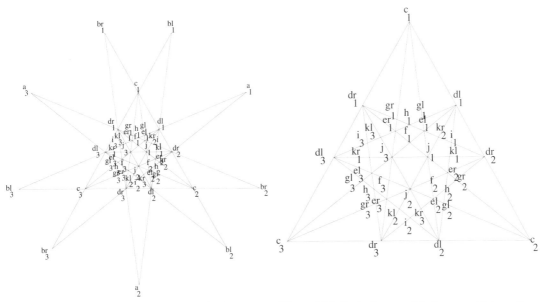

Figure 10.2–1: Intersections of all 20 planes   **Figure 10.2–2:** A closeup of the innermost intersections

computes the intersection of the lines *ab* and *cd*; unit[*vec*] computes a unit vector in the direction of *vec*. Listing 10.2–1 shows the code for defining all points *a* through *k*.

```
three = N[RotationMatrix2D[2Pi/3]]; (* a rotation by 120 degrees *)
inter[a_, b_, c_, d_] :=
   With[{p1 = a, v1 = unit[b-a], p2 = c, v2 = unit[c-d]},
     Module[{v1v2, s1, s2},
        v1v2 = v1[[1]] v2[[2]] - v1[[2]] v2[[1]];
        s1 = Det[{p2-p1, v2}]/v1v2;
        s2 = Det[{p2-p1, v1}]/v1v2;
        (p1 + v1 s1 + p2 + v2 s2)/2
     ] ]
norm[v_] := Sqrt[Plus@@(v^2)]
unit[v_] := v/norm[v]
define[s_Symbol, val1_] := (
        s[1] = N[val1];
        s[2] = three.s[1];
        s[3] = three.s[2]; )
define[c, {0, 1}] (* an equilateral triangle *)
define[ dl, c[1] + (1 - 1/GoldenRatio)(c[2] - c[1]) ]
define[ dr, c[1] + (1 - 1/GoldenRatio)(c[3] - c[1]) ]
define[ a,  inter[dl[3], dl[1], dr[3], dr[2]] ]
define[ bl, inter[ c[3],  c[1], dl[2], dl[1]] ]
define[ br, inter[ c[2],  c[1], dr[3], dr[1]] ]
define[ el, inter[ c[1], dl[2],  c[3], dl[1]] ]
define[ er, inter[ c[1], dr[3],  c[2], dr[1]] ]
define[ f,  inter[ c[2], dr[1],  c[3], dl[1]] ]
define[ gl, inter[ c[1], dl[2], dl[1], dl[3]] ]
define[ gr, inter[ c[1], dr[3], dr[1], dr[2]] ]
define[ h,  inter[dr[1], dr[2], dl[1], dl[3]] ]
define[ i,  inter[dl[1], dl[2], dr[1], dr[2]] ]
define[ j,  inter[ c[1], dl[2],  c[2], dr[1]] ]
define[ kl, inter[dl[3], dr[2], dl[1], dl[2]] ]
define[ kr, inter[dl[3], dr[2], dr[1], dr[3]] ]
```

Listing 10.2–1: Computation of the intersection points

The lines so constructed divide the plane into 15 different kinds of triangular or quadrilateral polygons, not counting the infinite exterior ones. The vertices of these polygons are the points *a* through *k*. Because the stellation should have icosahedral symmetry, the polygons in one plane that occur in this stellation must have threefold rotational symmetry. We combine sets of polygons with this symmetry into *facets*. The polygons in one facet are obtained from a single polygon by successive rotation by 120 degrees. Following Coxeter, we number the facets 0 through 14. Number 0 is the innermost triangle. The other facets have either threefold rotational or sixfold dihedral symmetry and consist of three polygons each. The *chiral* facets, that is, those with only

10.2 Stellations of the Icosahedron

rotational symmetry, come in sets of two, one set being the mirror image of the other one. These two sets are labeled 2_l and 2_r, for example. In some stellations, the two related chiral facets occur together. Their union can be treated like a single facet with dihedral symmetry. It is, therefore, denoted by 2 and consists of all six polygons from 2_l and 2_r.

Here are facets 0 through 13. Facet 14 is commonly regarded as part of 13, because the two facets always appear together, as we shall see.

The graphic object is the value of faceGraphicsInner. It uses the package Graphics'Legend' to produce the legend of color indices. The variable faceGraphicsInner contains a similar graphic object that leaves out the outermost facets, nos. 13 and 14. A color rendition of faceGraphicsInner is shown on Plate 2-a.

In[3]:= Show[faceGraphics];

Our notation for a facet with reflexive symmetry is facet[n]. Facets with only rotational symmetry (called *chiral* facets) are denoted by facet[n, R] and facet[n, L], respectively. Their union has reflexive symmetry and is denoted by facet[n].

These facets are purely formal objects. To define their geometry, we define numerical values for them. The definition

$$N[\text{facet}[2, R]] = \text{rot}[el[1], j[1], f[1]],$$

for example, says that the three polygons of facet[2, R] consist of the triangle $el_1 j_1 f_1$ and its two rotated forms. The function rot[*points*...] applies the rotations to the points given as the argument:

$$\text{rot[points\_\_]} := \text{NestList[(three.\#\& /@ \#)\&, \{points\}, 2]}.$$

Note the use of two nested pure functions. The variable three contains a matrix representing a

rotation by 120 degrees. The definition

$$N[\text{facet}[2]] = \text{merge}[2]$$

says that `facet[2]` consists of `facet[2, R]` and `facet[2, L]`.

Facet 0 is a single triangle.

```
In[4]:= N[ facet[0] ]
Out[4]= {{{0.126351, 0.072949}, {0., -0.145898},
        {-0.126351, 0.072949}}}
```

Here it is.

```
In[5]:= Show[
           Graphics[Polygon /@ %], AspectRatio -> Automatic
        ];
```

This chiral facet consists of three triangles.

```
In[6]:= N[ facet[2, L] ] // Short
Out[6]//Short=
  {{{-0.10222, 0.25}, {0., 0.17082}, {-0.126351, 0.072949}},
   <<1>>, {{-0.165396, -0.213525}, <<2>>}}

In[7]:= Show[
           Graphics[Polygon /@ %], AspectRatio -> Automatic
        ];
```

10.2 Stellations of the Icosahedron

10.2.2 Enumeration of All Possibilities

Our task is now to find all combinations of facets that lead to a valid stellation of the icosahedron. First, we have to define what we shall consider a valid stellation. Coxeter gives five conditions due to J. C. P. Miller:

1. The faces must lie in the 20 bounding planes of the icosahedron.

2. The parts of the faces in the 20 planes must be congruent. Those parts lying in one plane may be disconnected, however.

3. The parts lying in one plane must have threefold rotational symmetry, with or without reflections.

4. All parts must be accessible (they must lie on the outside of the solid).

5. Compounds that can be divided into two sets, each of which has the full symmetry of the whole, are excluded.

Condition 1 is automatically satisfied by our construction. Condition 3 guarantees that the stellation has icosahedral symmetry. Those stellations having the full icosahedral symmetry (including reflections) are called *reflexible*. Those that have only rotational symmetry are called *chiral*.

A stellation is described by a set of facets. Because facets have threefold rotational symmetry, this choice satisfies condition 3. This set of facets is the part of the face of the stellation that lies in one plane. By condition 2 this layout is the same in each of the 20 planes. Conditions 4 and 5 restrict the selection of subsets of the facets. All valid subsets can be found by geometrical reasoning. We will not repeat these considerations here but refer the reader to Coxeter's paper [11]. The stellation itself consists of 20 copies of the facets, one for each plane of the icosahedron.

The first stellation, the original icosahedron, for example, is {facet[0]}. facet[0] itself consists of a single triangle. The second stellation consists of facet 1, which is a set of three triangles. Facets 13 and 14 can only occur together. Therefore, we simply defined facet 13 as consisting of 13_l, 13_r, and the original 14.

It turns out that there are 59 possibilities. This explains the title of the paper mentioned: *The Fifty-Nine Icosahedra*. The enumeration is made simpler by noting that certain sets of facets are equivalent in the sense that either none or exactly one of them appears in a stellation. For chiral stellations, the sets

$$\lambda = \{3, 4\}$$
$$\mu = \{7, 8\}$$
$$\nu = \{11, 12\}$$

can be defined. The expression

$$\{\lambda, 5_r, 6_l, \mu, 9_l, 10_r, \nu\}$$

defines 6 solids, one for each choice of an element from the three sets λ, μ, and ν, together with facets 5_r, 6_l, 9_l, and 10_r.

These six stellations are nos. 54–59 in our list.

```
In[8]:= set6 = Icosahedra /@ Range[54, 59];
```

Here they are.

```
In[9]:= Show[ GraphicsArray[Partition[set6, 3]] ];
```

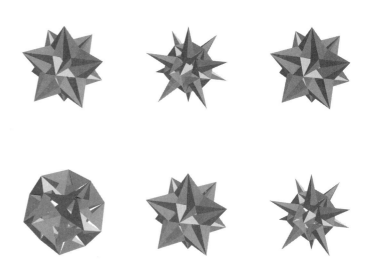

Similar sets can be defined for simplifying the enumeration of the reflexible stellations.

Choosing one element from each of the sets is best done with `Distribute[]`. The operation to distribute over is `List`. We show in a symbolic example how this works; see also Section 1.3.2.

Here are two sets with three and two elements, respectively.

```
In[10]:= alpha = {a1, a2, a3};\
         gamma = {g1, g2};
```

The distribution generates all six possible combinations.

```
In[11]:= Distribute[{alpha, b, gamma, d}, List]
Out[11]= {{a1, b, g1, d}, {a1, b, g2, d}, {a2, b, g1, d},
         {a2, b, g2, d}, {a3, b, g1, d}, {a3, b, g2, d}}
```

The result of the distribution is a list of stellations. They are then spliced into the main list `icosahedra` by turning the list into a sequence. The command `setOf[]` performs these manipulations. Listing 10.2–2 shows the code for producing all 59 stellations, excerpted from Icosahedra.m.

10.3 Rendering

```
(* sets of facets *)
lambda    = {facet[3], facet[4]}
lambdaref = {facet[3], facet[4], facet[5]}
mu        = {facet[7], facet[8]}
nu        = {facet[11], facet[12]}
nuref     = {facet[10], facet[11], facet[12]}
setOf[comps___] := Sequence @@ Distribute[ {comps}, List ]
icosahedra = {
  (* reflexible *)
  {facet[ 0]},
  {facet[ 1]},
  {facet[ 2]},
  {facet[13]},
  {facet[ 3], facet[ 4]},
  {facet[ 3], facet[ 5]},
  {facet[ 4], facet[ 5]},
  {facet[ 7], facet[ 8]},
  {facet[10], facet[11]},
  {facet[10], facet[12]},
  {facet[11], facet[12]},
  setOf[lambdaref, facet[6], mu],
  setOf[mu, facet[9], nuref],
  setOf[lambdaref, facet[6], facet[9], nuref],
  (* chiral *)
  {facet[ 5,L], facet[ 6,L], facet[ 9,R], facet[ 10,R]},
  setOf[lambda, facet[5,R], facet[6,L], facet[9,R], facet[10,R]],
  setOf[facet[5,R], facet[6,R], mu, facet[9,R], facet[10,R]],
  setOf[facet[5,R], facet[6,R], facet[9,L], facet[10,R], nu],
  setOf[lambda, facet[5,L], facet[6,R], mu, facet[9,R], facet[10,R]],
  setOf[lambda, facet[5,L], facet[6,R], facet[9,L], facet[10,R], nu],
  setOf[facet[5,L], facet[6,L], mu, facet[9,L], facet[10,R], nu],
  setOf[lambda, facet[5,R], facet[6,L], mu, facet[9,L], facet[10,R], nu]
};
```

Listing 10.2–2: Part of Icosahedra.m: All 59 stellations of the icosahedron

10.3 Rendering

So far, a stellation is a symbolic object, a set of facets. Because we already defined numerical values for the facets, it is now easy to turn them into a graphic object. The numerical value of a facet is a set of polygonal vertices that describes all parts of the stellation that lie in one plane. From this we can generate the *face graphic*, a construction plan for the solid. It can be used to assemble cardboard models by duplicating them 20 times, cutting out the pieces, and gluing them together.

Here, for example, is the planar map of stellation no. 59 (bottom right in Figure In[9]).

In[12]:= Show[FaceGraphics[59]];

To see the facets involved, you can consult the list icosahedra.

In[13]:= icosahedra[[59]]

Out[13]= {facet[4], facet[5, R], facet[6, L], facet[8], facet[9, L], facet[10, R], facet[12]}

10.3.1 Three-Dimensional Graphics

Instead of producing cardboard models, we shall now assemble the stellations into three-dimensional graphic objects. First, we turn the planar polygons into three-dimensional ones by inserting a z coordinate of 1 into each point. This prototypical face is then subjected to 20 (linear) maps in three dimensions to orient it into the 20 planes of the icosahedron. These 20 maps (represented as 3×3 matrices) are computed by finding the transformation that turns the points j_1, j_2, and j_3 of the prototypical face into the three vertices of one of the 20 faces of an icosahedron. We take the icosahedron from the package Graphics'Polyhedra'. The transformation is found by generating the nine equations for the coordinates and solving them with Solve[]. The details of the code can be found in the package Icosahedra.m.

The command Icosahedra[n] returns a Graphics3D object showing stellation no. n. An optional second argument of True reverses the object, if it is a chiral one. It has no effect for a reflexible stellation. Any graphic options given are passed along. Icosahedra[] itself takes two options, EdgeForm -> *list*, and PlaneColoring -> *func*. EdgeForm takes a list of graphics primitives, which are used as the argument of the graphic *primitive* EdgeForm[] to control the rendering of the edges. Because a planar face may be composed of many parts, it is usually best to use the default {}, which suppresses edges. The option PlaneColoring takes a function as value. This function will be called with an integer argument between 1 and 20 and should return the color directives for rendering the ith plane. (Planes are ordered such that parallel planes are adjacent to each other.) The default uses 10 different hues, one for each pair of parallel planes (20 different hues cannot be distinguished easily on most output devices). It is visible only if lighting is turned off. Here are a few examples of using these commands and options.

10.3 Rendering

Here are the right and left forms of the compound of five tetrahedra.

```
In[14]:= Show[ GraphicsArray[
                {Icosahedra[36], Icosahedra[36, True]}] ];
```

The two forms can be combined to give the compound of ten tetrahedra. It is no. 18 in our list.

```
In[15]:= Show[ Icosahedra[18] ];
```

Here is the compound of five octahedra with unique colors for the different planes. A color rendering is on Plate 2-b.

```
In[16]:= Show[ Icosahedra[3], Lighting -> False ];
```

The pattern used in `Icosahedra[]` for the optional second argument is worth mentioning. It should accept either `True` or `False` and should have a default of `False`. Here is the left side of the definition of `Icosahedra[]` with this pattern:

```
Icosahedra[n_, rev:(True|False):False, opts___] := ....
```

The two colons mean different things: the first one declares `rev` as a pattern variable, the second one introduces a default. The pattern itself uses an alternative for the two cases. `FaceGraphics[]` also allows this optional argument. The package contains a few additional commands not mentioned here. Reading it in takes some time, because many properties are computed at that time, for example the 20 linear maps described earlier.

10.3.2 Ordering by Increasing Size

The order of the 59 stellations in our list `icosahedra` is determined by combinatorial aspects (see Listing 10.2–2). Often, an ordering according to the size (circumradius) is desired. We can order the icosahedra in this way by defining a suitable ordering predicate. First, we order the facets according to increasing radius. The sorted list is assigned to the variable `facetorder` (see Listing 10.3–1). The function `sizedFacets[i]` gives the list of facets of stellation no. i, in order of increasing size. An inverse lexicographic ordering on such lists of faces is our desired ordering.

```
facetorder = {0, 1, 2, 3, 4, 6, 7, 5, 9, 10, 12, 8, 11, 13}
sizeRules = Thread[facetorder -> Range[0, 13]]
sizedFacets[i_Integer] := Sort[ icosahedra[[i]] /. facet[n_, ___] :> n /. sizeRules ]
icosaOrderedQ[i1_, i2_] := invlex[ sizedFacets[i1], sizedFacets[i2] ]
invlex[{}, _] := True
invlex[{___, l1_}, {___, l2_}] /; l1 < l2 := True
invlex[{r1___, l_}, {r2___, l_}] := invlex[{r1}, {r2}]
invlex[_, _] := False
sizedIcosahedra := sizedIcosahedra = Sort[ Range[Length[icosahedra]], icosaOrderedQ ]
```

Listing 10.3–1: Part of Icosahedra.m: Sorting by increasing size

Here are the numbers of the stellations in increasing size.

```
In[17]:= sizedIcosahedra
Out[17]= {1, 2, 3, 5, 12, 14, 6, 7, 16, 24, 27, 18, 30, 33,
          34, 35, 36, 40, 42, 26, 29, 20, 32, 10, 39, 45, 47, 49,
          53, 57, 13, 15, 8, 17, 21, 37, 41, 43, 23, 51, 55, 59,
          25, 28, 19, 31, 9, 38, 44, 46, 48, 52, 56, 11, 22, 50,
          54, 58, 4}
```

Inspecting the sorted list, we find that the size of the stellation increases at these positions.

```
In[18]:= jumps = {1, 2, 3, 4, 5, 7, 10, 20, 31, 43, 59};
```

10.3 Rendering

Here are the corresponding numbers of the icosahedra in the original order.

```
In[19]:= origs = sizedIcosahedra[[jumps]]
Out[19]= {1, 2, 3, 5, 12, 6, 24, 26, 13, 25, 4}
```

We partition them into a two-dimensional array.

```
In[20]:= Join[ Partition[ origs, 4 ],
               {Take[origs, -Mod[Length[origs], 4]]} ]
Out[20]= {{1, 2, 3, 5}, {12, 6, 24, 26}, {13, 25, 4}}
```

We determine the plot range needed for the largest stellation.

```
In[21]:= FullOptions[ Icosahedra[origs[[-1]]], PlotRange ];
```

We generate the graphics, fixing the plot range to the same value.

```
In[22]:= Map[ Icosahedra[#, PlotRange -> %]&, %%, {2} ];
```

Figure 10.3–1 shows the true size relations.

```
In[23]:= Show[ GraphicsArray[%] ];
```

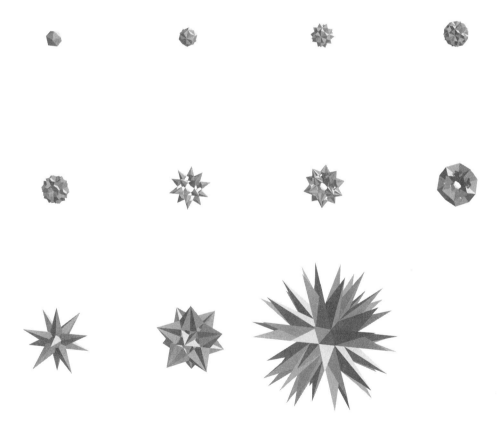

Figure 10.3–1: The true size relations of the stellated icosahedra

10.4 Discussion

The five octahedra circumscribed about an icosahedron were first mentioned around 1900. (Coxeter's paper [11] gives detailed references.) From it, the compounds of 5 and 10 tetrahedra can be deduced. A total of 22 stellations were described by Brückner [9] and Wheeler [59].

This command draws the three-dimensional picture and the planar map next to each other. The result is an image that is similar to the plates in Coxeter's paper. You can find such images of all stellations in the notebook Icosahedra.nb.

```
In[24]:= icosaFigure[n_, opts___] :=
            Show[ GraphicsArray[ {Icosahedra[n, opts],
                  FaceGraphics[n, opts, Frame -> True,
                               FrameTicks -> None]} ]
                ]
```

The first stellation is formed by facet 2. It is similar to the figure used at one point by Wolfram Research as its logo (and still used as a *Mathematica* icon).

```
In[25]:= icosaFigure[ 2 ];
```

The icon proper was wrongly called the "stellated icosahedron." It should be called a *faceted icosahedron*.

```
In[26]:= Show[ Stellate[Polyhedron[Icosahedron]] ];
```

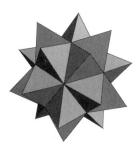

10.4 Discussion

Here is the great icosahedron, no. 11 in our list. In contrast to the version from Graphics`Polyhedra`, this construction generates only the visible parts of the faces and is, therefore, easier to render because no intersections of polygons have to be computed. See Chapter 9 for another approach to rendering this uniform polyhedron.

In[27]:= icosaFigure[11];

Here is the largest of the stellations. It contains facet 13 (which includes 14, as explained earlier).

In[28]:= icosaFigure[4];

No. 8 is the only stellation whose solid pieces are completely disconnected (they do not even share vertices).

In[29]:= icosaFigure[8];

Here is the prototypical chiral stellation. Every chiral stellation is obtained by adding these facets (or their reflections) to a reflexible stellation. Note that it is only vertex connected.

In[30]:= icosaFigure[33];

This (reflexible) stellation is the only one that consists entirely of chiral parts. It is the combination of no. 33 (shown in the previous picture) and its mirror image.

In[31]:= icosaFigure[30];

Plate 6 shows all 59 stellations, rendered with ray tracing (see Chapter 11). They have been arranged by increasing size across the two pages, as discussed in Section 10.3.2.

 The CD-ROM contains the package Icosahedra.m, as well as the notebook Icosahedra.nb, with the examples from this chapter.

10.5   The Complete Code of Icosahedra.m

```
BeginPackage["MathProg`Icosahedra`"]

Icosahedra::usage = "Icosahedra[n, opts...] is a 3D graphics object
    representing stellation no. n of the icosahedron."

EdgeForm::usage = EdgeForm::usage <> "EdgeForm -> {primitives} is an
    option of Icosahedra that gives the rendering of edges."
PlaneColoring::usage = "PlaneColoring->colorfunction is an option of Icosahedra that
    gives the color function of each planar face.
    Argument is a number from 1 to 20."
lineGraphics::usage = "lineGraphics is a graphics object showing
    the intersection of one facial plane with all others."
lineGraphicsInner::usage = "lineGraphicsInner leaves out the outermost
    points of lineGraphics."
faceGraphics::usage = "faceGraphics is a graphics of color coded facets."
faceGraphicsInner::usage = "faceGraphicsInner leaves out the outermost
    facets of faceGraphics."
facet::usage = "facet[n] or facet[n, r|l] is a symbolic representation
    of one facet. N[facet[...]] gives the vertex coordinates."
icosahedra::usage = "icosahedra is the list of all stellated icosahedra in symbolic form."
sizedIcosahedra::usage = "sizedIcosahedra is the list of stellations
    sorted by increasing size."
FaceGraphics::usage = "FaceGraphics[n, (True|False)] is
    one planar face of stellation no. n."
faces::usage = "faces[n, (True|False)] is a numerical representation
    of one planar face of stellation no. n."
r::usage = "r designates the dextro form of a chiral object."
l::usage = "l designates the laevo form of a chiral object."

(* points *)
{a, bl, br, c, dl, dr, el, er, f, gl, gr, h, i, j, kl, kr}

Begin["`Private`"]

Needs["Graphics`Polyhedra`"]
Needs["Utilities`FilterOptions`"]
Needs["Geometry`Rotations`"]
Needs["Graphics`Legend`"]

(* points in one plane *)

prec = 30;
n[a_] := N[a, prec] (* higher precision for intermediate computations *)

three = n[RotationMatrix2D[2Pi/3]]; (* a rotation by 120 degrees *)

(* intersection of lines ab and cd in two dimensions *)
```

```
inter[a_, b_, c_, d_] :=
    With[{p1 = a, v1 = unit[b-a], p2 = c, v2 = unit[c-d]},
     Module[{v1v2, s1, s2},
     v1v2 = v1[[1]] v2[[2]] - v1[[2]] v2[[1]];
     s1 = Det[{p2-p1, v2}]/v1v2;
     s2 = Det[{p2-p1, v1}]/v1v2;
     (p1 + v1 s1 + p2 + v2 s2)/2
      ]
     ]
norm[v_] := Sqrt[Plus@@(v^2)]
unit[v_] := v/norm[v]

(* make three definitions *)

define[s_Symbol, val1_] := (
    s[1] = n[val1];
    s[2] = three.s[1];
    s[3] = three.s[2];
)

(* points *)

define[c, {0, 1}]

define[ dl, c[1] + (1 - 1/GoldenRatio)(c[2] - c[1]) ]
define[ dr, c[1] + (1 - 1/GoldenRatio)(c[3] - c[1]) ]

define[  a, inter[dl[3], dl[1], dr[3], dr[2]] ]

define[ bl, inter[c[3], c[1], dl[2], dl[1]] ]
define[ br, inter[c[2], c[1], dr[3], dr[1]] ]

define[ el, inter[c[1], dl[2], c[3], dl[1]] ]
define[ er, inter[c[1], dr[3], c[2], dr[1]] ]

define[  f, inter[c[2], dr[1], c[3], dl[1]] ]

define[ gl, inter[c[1], dl[2], dl[1], dl[3]] ]
define[ gr, inter[c[1], dr[3], dr[1], dr[2]] ]

define[  h, inter[dr[1], dr[2], dl[1], dl[3]] ]

define[  i, inter[dl[1], dl[2], dr[1], dr[2]] ]

define[  j, inter[c[1], dl[2], c[2], dr[1]] ]

define[ kl, inter[dl[3], dr[2], dl[1], dl[2]] ]
define[ kr, inter[dl[3], dr[2], dr[1], dr[3]] ]

(* line drawings *)

mod[n_] := Mod[n-1, 3] + 1

line[a_, b_] := Line[{a[#], b[#]}]& /@ {1, 2, 3}
line[a_, b_, d_] := Line[{a[#], b[mod[#+d]]}]& /@ {1, 2, 3}

point[s_Symbol] :=
    Text[HoldForm[Subscripted[s[#]]], s[#], -1.1 unit[s[#]]]& /@ {1, 2, 3}
```

Icosahedra.m   229

```
lineGraphics := lineGraphics =
Graphics[ {
    Thickness[0],
    point[c],
    point[dl], point[dr],
    line[c, dl, 1],
    line[c, dr, 2],
    point[el], point[er],
    point[f],
    point[gl], point[gr],
    point[h],
    point[i],
    point[j],
    point[kl], point[kr],
    line[dr, dl, 1],
    point[a],
    point[bl], point[br],
    line[bl, a, 1], line[br, a, 1],
    line[bl, br, 2]
    }, {
    AspectRatio->Automatic, PlotRange->All,
    DefaultFont -> {"Times-Roman", 6.0}
    } ]
lineGraphicsInner := lineGraphicsInner =
Graphics[ {
    Thickness[0],
    point[c],
    point[dl], point[dr],
    line[c, dl, 1],
    line[c, dr, 2],
    point[el], point[er],
    point[f],
    point[gl], point[gr],
    point[h],
    point[i],
    point[j],
    point[kl], point[kr],
    line[dr, dl, 1],
    line[dl, dl, 1], line[dr, dr, 2],
    line[c, c, 2]
    }, {
    AspectRatio->Automatic, PlotRange->All,
    DefaultFont -> {"Times-Roman", 8.0}
    } ]
(* Coxeter faces *)
(* generate three polygons out of one by rotation *)
rot[points__] := NestList[(three.#& /@ #)&, {points}, 2]
(* combine the left and right facets *)
merge[n_] := Join[ N[facet[n,r]], N[facet[n,l]] ]
```

```
N[facet[0]] = {{j[1], j[2], j[3]}}; (* this one has only one piece *)
N[facet[1]] = rot[j[1], j[3], f[1]]
N[facet[2, r]] = rot[el[1], j[1], f[1]]
N[facet[2, l]] = rot[er[1], f[1], j[3]]
N[facet[2]] = merge[2]
N[facet[3]] = rot[h[1], el[1], f[1], er[1]]
N[facet[4, r]] = rot[el[1], kr[2], j[1]]
N[facet[4, l]] = rot[er[1], j[3], kl[3]]
N[facet[4]] = merge[4]
N[facet[5, r]] = rot[dl[1], kr[2], el[1]]
N[facet[5, l]] = rot[dr[1], er[1], kl[3]]
N[facet[5]] = merge[5]
N[facet[6, r]] = rot[gl[1], el[1], h[1]]
N[facet[6, l]] = rot[gr[1], h[1], er[1]]
N[facet[6]] = merge[6]
N[facet[7]] = rot[i[1], kl[1], j[1], kr[2]]
N[facet[8]] = rot[c[1], gl[1], h[1], gr[1]]
N[facet[9, r]] = rot[dl[1], i[1], kr[2]]
N[facet[9, l]] = rot[dr[1], kl[3], i[3]]
N[facet[9]] = merge[9]
N[facet[10, r]] = rot[dl[1], el[1], gl[1]]
N[facet[10, l]] = rot[dr[1], gr[1], er[1]]
N[facet[10]] = merge[10]
N[facet[11, r]] = rot[c[1], dl[1], gl[1]]
N[facet[11, l]] = rot[c[1], gr[1], dr[1]]
N[facet[11]] = merge[11]
N[facet[12]] = rot[dl[1], dr[2], i[1]]
N[facet[13, r]] = rot[bl[1], dl[1], c[1]]
N[facet[13, l]] = rot[br[1], c[1], dr[1]]
N[facet[14]] = rot[a[1], dr[2], dl[1]]
N[facet[13]] = Join[ merge[13], N[facet[14]] ] (* ! 14 is always with 13 *)
(* facet color, n=0,..,max *)
facetColor[n_, max_] := Hue[n/(max+1),1,1]
(* color coded map *)
colorMap[max_] := Graphics[
    {facetColor[#, max], Polygon /@ N[facet[#]]}& /@ Range[0, max],
    {AspectRatio->Automatic, PlotRange->All}
]
legend[max_] := {
    {facetColor[#, max],
     ToString[NumberForm[#, 2, NumberPadding->" "]]}& /@ Range[0, max],
    LegendPosition->{-1, 0}
}
```

Icosahedra.m

```
faceGraphics := faceGraphics =
    Block[{$DisplayFunction=Identity},
        ShowLegend[ colorMap[13], legend[13] ]
    ]
faceGraphicsInner := faceGraphicsInner =
    Block[{$DisplayFunction=Identity},
        ShowLegend[ colorMap[12], legend[12] ]
    ]
(* sets of facets *)
lambda = {facet[3], facet[4]}
lambdaref = {facet[3], facet[4], facet[5]}
mu = {facet[7], facet[8]}
nu = {facet[11], facet[12]}
nuref = {facet[10], facet[11], facet[12]}
setOf[comps___] := Sequence @@ Distribute[ {comps}, List ]
icosahedra = {
  (* reflexible *)
    {facet[ 0]},
    {facet[ 1]},
    {facet[ 2]},
    {facet[13]},
    {facet[ 3], facet[ 4]},
    {facet[ 3], facet[ 5]},
    {facet[ 4], facet[ 5]},
    {facet[ 7], facet[ 8]},
    {facet[10], facet[11]},
    {facet[10], facet[12]},
    {facet[11], facet[12]},
    setOf[lambdaref, facet[6], mu],
    setOf[mu, facet[9], nuref],
    setOf[lambdaref, facet[6], facet[9], nuref],
  (* chiral *)
    {facet[ 5,l], facet[ 6,l], facet[ 9,r], facet[ 10,r]},
    setOf[lambda, facet[5,r], facet[6,l], facet[9,r], facet[10,r]],
    setOf[facet[5,r], facet[6,r], mu, facet[9,r], facet[10,r]],
    setOf[facet[5,r], facet[6,r], facet[9,l], facet[10,r], nu],
    setOf[lambda, facet[5,l], facet[6,r], mu, facet[9,r], facet[10,r]],
    setOf[lambda, facet[5,l], facet[6,r], facet[9,l], facet[10,r], nu],
    setOf[facet[5,l], facet[6,l], mu, facet[9,l], facet[10,r], nu],
    setOf[lambda, facet[5,r], facet[6,l], mu, facet[9,l], facet[10,r], nu]
};
(* lists of faces. A face is a list of vertices.
   A vertex is a list of 2 numbers *)
faces[n_, rev_:False] := Join @@ N[icosahedra[[n]]]
faces[n_, True] :=
    Join @@ N[ icosahedra[[n]] /.
              {facet[i_, r] :> facet[i, l], facet[i_, l] :> facet[i, r]} ]
```

```
faces3D[n_, rev_:False] :=
    Apply[ {#1, #2, 1}&, faces[n, rev], {2} ] (* z coordinate is 1 *)
FaceGraphics[n_Integer, rev:(True|False):False, opts___] :=
    Graphics[ Polygon /@ faces[n, rev], {opts, AspectRatio->Automatic} ]
(* compute plane transforms *)
(* we take an ordinary icosahedron as template *)
targetfaces = Polyhedron[Icosahedron][[1]] /. Polygon[f_] :> f
(* compute normal vectors *)
cross[ {ax_, ay_, az_}, {bx_, by_, bz_} ] :=
    {ay bz - az by, az bx - ax bz, ax by - ay bx}
normal[f_List] := unit[cross[ f[[2]] - f[[1]], f[[3]] - f[[2]] ]]
normals = normal /@ targetfaces;
(* find opposite pairs *)
perm = Flatten[ Position[ normals, #, {1} ]& /@ -normals ]
pairs = Transpose[{Range[Length[perm]], perm}]
pairs = Flatten[ Union[Sort /@ pairs] ]
If[ Union[pairs] === Range[Length[targetfaces]], (* sanity check *)
    targetfaces = targetfaces[[Flatten[pairs]]]
]

(* find the matrix that transforms the three points in template
   into the three points in goal *)
trans[template_, goal_] :=
    Module[{mat = Array[m, {3, 3}]},
        eqs = mat . Transpose[template] == Transpose[goal];
        eqs = Thread[ Flatten /@ eqs ];
        sol = Solve[ eqs, Flatten[mat] ];
        mat /. sol[[1]]
    ]
(* facet[0] should map into the icosahedron faces *)
proto = faces3D[1][[1]];
(* a list of 20 transformation matrices *)
transforms = trans[proto, #]& /@ targetfaces;
(* graphics *)
Options[Icosahedra] = {
    EdgeForm -> {},
    PlaneColoring -> (Hue[Floor[(#-1)/2]/10]&)
}
defaults = {PlotRange->All}; (* changes to standard options *)
```

Icosahedra.m

```
Icosahedra[n_, rev:(True|False):False, opts___] /; 1 <= n <= Length[icosahedra] :=
    With[{plane = faces3D[n, rev]},
    Module[{planes, edges, colorfunction},
        planes = Function[trans, Map[trans . #&, plane, {2}]] /@ transforms;
        planes = Map[Polygon, planes, {2}];
        edges = EdgeForm /. {opts} /. Options[Icosahedra];
        colorfunction = PlaneColoring /. {opts} /. Options[Icosahedra];
        planes = MapIndexed[ Prepend[#1, colorfunction[#2[[1]]]]&, planes ];
        Graphics3D[ Prepend[planes, EdgeForm[edges]],
                    Join[{FilterOptions[Graphics3D, opts]}, defaults] ]
    ]]
(* sorting by size *)
facetorder = {0, 1, 2, 3, 4, 6, 7, 5, 9, 10, 12, 8, 11, 13}
sizeRules = Thread[facetorder -> Range[0, 13]]
sizedFacets[i_Integer] := Sort[ icosahedra[[i]] /. facet[n_, ___] :> n /. sizeRules ]
icosaOrderedQ[i1_, i2_] := invlex[ sizedFacets[i1], sizedFacets[i2] ]
invlex[{}, _] := True
invlex[{___, l1_}, {___, l2_}] /; l1 < l2 := True
invlex[{r1___, l_}, {r2___, l_}] := invlex[{r1}, {r2}]
invlex[_, _] := False
sizedIcosahedra := sizedIcosahedra =
    Sort[ Range[Length[icosahedra]], icosaOrderedQ ]
End[]
Share[]
Protect[ Icosahedra, facet, icosahedra, FaceGraphics, faces, r, l ]
EndPackage[]
```

Listing 10.5–1: Icosahedra.m

Chapter Eleven

Ray Tracing

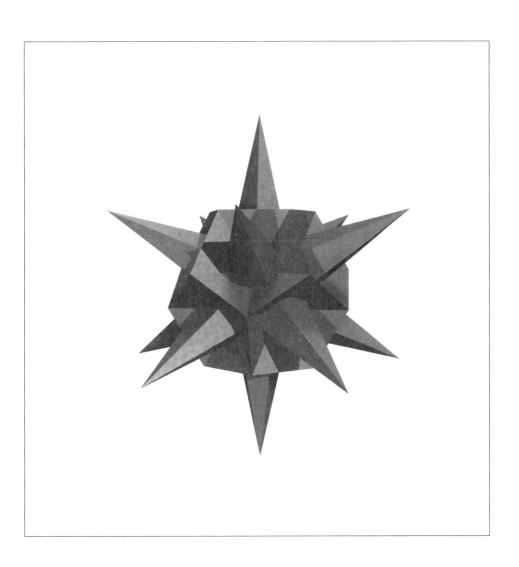

In Section 1, we develop a data type for surfaces in space that maintains important connectivity data and surface normals. The new data type is especially important for preparing ray-tracing input from *Mathematica* graphics. Section 2 discusses ray tracing and Section 3 treats conversion utilities for preparing input from *Mathematica* graphics for two popular ray-tracing programs, POVRAY and RAYSHADE. Section 4 shows how the color plates were generated. Finally, Section 5 shown an application: ray-traced stereo pairs of images that can give a true three-dimensional appearance.

Ray tracers make it feasible to produce stunning images with moderate effort. The graphics conversion programs described here allow you to tap *Mathematica*'s potential to easily create input for ray tracing.

About the illustration overleaf:
The stellated icosahedron from Chapter 10 in a ray-traced version.

The code for this picture is in BookPictures.m.

11.1 A Data Type for Surfaces

Surfaces in space are usually represented by a collection of polygons—most often triangles or quadrilaterals. A simple rendering technique for such (flat) polygons gives each one a single color, determined by the reflection properties of the light sources in the scene. The reflection properties are computed with the help of the surface normal direction. Because each polygon has its own distinct color, the surface patches (the polygons) are clearly visible. *Mathematica*'s POSTSCRIPT renderer uses this technique.

Parametric surfaces are rendered in the way just described. A color version of this torus is shown in Plate 7-a.

```
In[1]:= ParametricPlot3D[ {Cos[phi](2+Cos[psi]),
                          Sin[phi](2+Cos[psi]), Sin[psi]},
         {phi, 0, 2Pi}, {psi, 0, 2Pi},
         PlotPoints->{36, 18}, Axes->None, Boxed->False
       ];
```

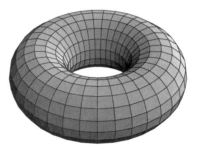

The internal data structure is a list of quadrilaterals (that is, nonplanar polygons with four edges).

```
In[2]:= Short[ %[[1]] ]
Out[2]//Short=
   {Polygon[{{3., 0., 0.}, {2.95179, 0.535671, 0.}, <<1>>,
     {2.93247, 0., 0.361242}}], <<594>>}
```

For a more realistic rendering, we try to hide the visible internal edges to produce a smooth surface. Smoothing is achieved by interpolating color values within the polygons. Instead of computing a single surface normal for a polygon, we give each vertex a normal direction and the normal direction inside the polygon is interpolated from the values at the corners. This interpolation can be performed in real time on modern graphics workstations. It is the main technique that popular scientific visualization programs such as AVS or Geomview use to achieve good-looking pictures.

Because each vertex appears in four different polygons, we must make sure that all four instances of a vertex are given the same normal direction. Our data structure makes it difficult to achieve this, because the connectivity information has been lost. If we want to use *Mathematica* to generate data that can be input into external rendering programs, we need to conserve the connectivity data. Moreover, due to the symbolic nature of *Mathematica*, we can compute surface normal vectors exactly, by symbolic differentiation!

The basic idea is simple. If we replace the command `ParametricPlot3D` by `Table` we get a matrix of points. A similar matrix of normal vectors can be computed by forming the cross product of the two partial derivatives. The resulting data structure contains all information to drive external renderers or to generate polygons for *Mathematica*'s own renderer. This idea is implemented by defining a new graphic data type

$$\texttt{SurfaceGraphics3D[}\textit{vertices, normals, }\{\textit{options}\ldots\}\texttt{]}\ .$$

To allow for cases where the normals cannot be computed, the second element is optional. The last element stores the options given when the graphic data was produced. The built-in types, such as `Graphics3D` and `SurfaceGraphics`, work in the same way.

Our main source of `SurfaceGraphics3D` is `ParametricPlot3D`. We override the built-in definition to produce our new graphic data type instead of `Graphics3D`. The code is in the package SurfaceGraphics3D.m, reproduced in Listing 11.1–1. The code in the auxiliary procedure `MakeSurface3D` generates the table of vertices and normals. A test is made to see whether the normals are numeric. If they are not, the normals are not included in the output. Normals are not computed, either, if the parametric expression f is not given explicitly as a list of three expressions. (Often, `ParametricPlot3D[Evaluate[`f`], ...]` can be used in such cases.)

The code for `ParametricPlot3D` performs the calculations to determine the number of plot points in each direction in the same way that the built-in code does. The package can be used together with the graphic package Graphics'ParametricPlot3D', which allows alternate ways to specify the number of plot points and which contains additional commands for spherical and cylindrical plots.

Naturally, we want to display this new graphic data type with *Mathematica*'s renderer, as well. The easiest way to do so is to convert it back into a `Graphics3D` object. The conversion is implemented as a type conversion of the form `Graphics3D[ SurfaceGraphics3D[...] ]`, which is the standard way of converting one graphic form into another one. Converting a matrix of points into a set of quadrilaterals by picking each possible choice of matrix elements a_{ij}, $a_{i\,j+1}$, $a_{i+1\,j+1}$, and $a_{i+1\,j}$ is a nice programming exercise. Rules (upvalues) for `Show` and `Display[]` and a special format help to integrate the new data type seamlessly into *Mathematica*.

The new data type can be loaded at any time.	`In[3]:= Needs["MathProg`SurfaceGraphics3D`"]`

The same command for the torus returns now a `SurfaceGraphics3D` object. When it is shown with `Show[%]`, it generates the same picture as before; it is, therefore, not drawn here.	`In[4]:= ParametricPlot3D[ {Cos[phi](2+Cos[psi]),`
	`                Sin[phi](2+Cos[psi]), Sin[psi]},`
	`       {phi, 0, 2Pi}, {psi, 0, 2Pi},`
	`       PlotPoints->{24, 12}, Axes->None, Boxed->False`
	`     ]`
	`Out[4]= -SurfaceGraphics3D-`

11.1 A Data Type for Surfaces

The data type contains the vertices, normal vectors, and options.

```
In[5]:= Short[InputForm[%], 3]
Out[5]//Short=
 SurfaceGraphics3D[{<<12>>},
   {{{-1., 0, 0}, <<22>>, {-1., 0, 0}}, <<11>>},
   {PlotPoints -> {24, 12}, Axes -> None, Boxed -> False}]
```

So far, we have not gained anything, because we could not take advantage of the extra information available in `SurfaceGraphics3D`. Therefore, let us now explore external renderers.

```
BeginPackage["MathProg`SurfaceGraphics3D`"]

SurfaceGraphics3D::usage = "SurfaceGraphics3D[vertices, (normals,) {opts...}]
    represents a surface with given vertices and normals."

Begin["`Private`"]

Needs["Utilities`FilterOptions`"]

protected = Unprotect[ParametricPlot3D]

(* overload ParametricPlot3D[] to produce SurfaceGraphics3D *)

ParametricPlot3D[f_, {u_, u0_, u1_}, {v_, v0_, v1_}, opts___] :=
    Module[{plotpoints, ndu, ndv},
        plotpoints = PlotPoints /. {opts} /. Options[ParametricPlot3D];
        If[ Head[plotpoints] =!= List, plotpoints = {plotpoints, plotpoints} ];
        plotpoints = plotpoints /. Automatic -> 15;
        ndu = (u1-u0)/(plotpoints[[1]]-1);
        ndv = (v1-v0)/(plotpoints[[2]]-1);
        MakeSurface3D[f, {u, u0, u1, ndu}, {v, v0, v1, ndv}, opts]
    ]

(* compute normals *)

MakeSurface3D[f:{_, _, _}, ur:{u_, _, _, _}, vr:{v_, _, _, _}, opts___ ] :=
    Module[{n, coords, surf, normals},
        n = Cross[D[f, v], D[f, u]];    (* normal vectors *)
        coords = Table[ N[{f, unit[n]}], ur, vr ];
        {surf, normals} = Transpose[ coords, {3, 2, 1} ];
        If[ And @@ NumberQ /@ Flatten[normals],
            SurfaceGraphics3D[ surf, normals, {opts} ],
        (* else: couldn't compute normals *)
            SurfaceGraphics3D[ surf, {opts} ]
        ]
    ]

unit[v_] := v/Sqrt[Plus@@(v^2)]

(* strange functions. Do not attempt to produce normals *)

MakeSurface3D[f_, ur:{u_, _, _, _}, vr:{v_, _, _, _}, opts___ ] :=
        SurfaceGraphics3D[ Table[N[f], ur, vr], {opts} ]

(* convert to Graphics3D *)
```

```
makeMesh[vl_List] :=
    Module[{l = Drop[#, -1]& /@ vl, ll = Drop[#, 1]& /@ vl, mesh},
        mesh = {Drop[l, -1], Drop[ll, -1], Drop[ll, 1], Drop[l, 1]};
        Transpose[ Flatten[#, 1]& /@ mesh ]
    ]
SurfaceGraphics3D/:
Graphics3D[SurfaceGraphics3D[vl_List, normals_:{}, {opts___}]] :=
    Graphics3D[ Polygon /@ makeMesh[vl], {FilterOptions[Graphics3D, opts]} ]
SurfaceGraphics3D/:
Show[s_SurfaceGraphics3D, opts___] := Show[Graphics3D[s], opts]
SurfaceGraphics3D/:
Display[file_, s_SurfaceGraphics3D, args___] :=
    Display[file, Graphics3D[s], args]
Format[_SurfaceGraphics3D] = "-SurfaceGraphics3D-"
Protect[ Evaluate[protected] ]
End[]

Protect[ SurfaceGraphics3D ]
EndPackage[]
```

Listing 11.1–1: SurfaceGraphics3D.m

11.2 Photorealistic Rendering

Ray tracing is one method for photorealistic rendering. As the name suggests, light rays are followed from their origin at a light source through the scene and eventually into the eye or camera. For efficiency reasons, rays are usually traced in the opposite direction, starting at the camera. When a ray intersects an object, shadow rays are generated toward each light source to find out which light sources contribute to the illumination at this point, and which ones are in the shadow, because they are blocked by other objects. If the surface is reflective (like a mirror), additional reflected rays are generated and followed in the same way. If the surface is transparent, refracted rays are generated. The illumination at a point of intersection is determined from the material properties and the angles between the rays and the light sources. A scene description for ray tracing consists of these parts:

- The properties of the camera: position, orientation, focal width, and resolution (pixels horizontal and vertical)

- The properties of the light sources: position, direction (for spot lights), extension (for sources other than point lights), and color of the light

- Objects, composed of graphic primitives, such as spheres, planes, triangles
- Material properties of objects, such as reflectivity, shininess, color, grainyness
- Textures, giving more realistic surfaces.

The torus from Plate 7-a, together with a simple camera and a few point light sources, produces the scene description given in Listing 11.2–1. The syntax used is that of the ray-tracing program POVRAY. The list of triangles was generated from *Mathematica*, the tools for which are described in the next section. The resulting picture is shown in Plate 7-b.

If we use our data type for surfaces, we can include the normal vectors in the scene description. Instead of generating `triangle` objects, we can produce `smooth_triangle` objects. The program uses *Phong shading* to interpolate the normal direction inside the triangles. The result is shown in Plate 7-c.

Ray tracing does not take into account all interactions of objects and light rays. While it is very good at computing shadows and reflections or refractions of objects, it does not account for reflected light rays. Light sources—like vampires—do not reflect in mirrors. Nor do lights reflect off diffuse objects. This deficiency makes it difficult to render ambient light, which is so important for real-life scenes. A more sophisticated method, called *radiosity*, has been developed for taking into account these phenomena. It is much more computer intensive than is ray tracing. With experimentation and experience, ray tracing can be used to generate realistic scenes, and it has become quite popular.

Plate 7-d shows some of the effects that can be achieved, using the same torus data as Plate 7-c. The textures and surface properties used here are predefined in an include file that is part of the POVRAY software. The floor shows a checkerboard texture, the rear wall has a grainy finish, a mirror has been placed in front of this wall. The torus has a granite texture. Observe the shadows and highlights present. For more information about ray tracing, refer to the literature, for example, [22].

The amount of realism added to a scene influences rendering time. Plates 7-b through 7-d were all rendered at a resolution of 840×630 pixels. Plate 7-b took 5 minutes, whereas Plate 7-d took 35 minutes on a SPARCstation 5.

11.3 Converting *Mathematica* Graphics

External renderers, such as AVS, Geomview, or ray-tracing programs all need their input in a certain data format. Few standards exist, and each program invents its own format. The level of description of geometrical objects is usually rather low. Points are the primitive data type. On the other hand, *Mathematica*'s graphic operations are rather high level. You can use formulae to describe mathematical or geometrical objects. Because it is rather easy to program *Mathematica*

```
camera {
    location <4.0899, -7.5424, 3>
    direction <0, 4.5, 0>
    up <0, 0, 3.>
    right <4., 0, 0>
    sky <0., 0., 1.>
    look_at <0, 0, 0.1>
}
light_source {   <0,0,5>
    color rgb <0.9, 0.9, 0.7>
}
light_source {   <5, 4, 1.5>
    color rgb <0.8, 0.8, 0.8>
}
light_source {   <3, -8, -0.5>
    color rgb <1, 0.8, 0.8>
}
#default {
    texture { pigment {color rgb <1,1,1>}
         finish {phong 0.8 ambient 0.3}
    }
}
/* begin Graphics3D */
triangle {
    <3., 0, 0>
    <2.8888, 0.80939, 0>
    <2.7359, 0.76656, 0.54064> }
triangle {
    <2.7359, 0.76656, 0.54064>
    <2.8413, 0, 0.54064>
    <3., 0, 0> }
⋮
triangle {
    <3., 0, 0>
    <2.8888, -0.80939, 0>
    <2.7359, -0.76656, -0.54064> }
/* end Graphics3D */
```

Listing 11.2–1: Part of figure2.pov, the input to POVRAY for the torus

to output its data in a variety of formats, it is the ideal tool for generating the data and then converting it into the form required by the ray-tracing program used. This technique was used very early in the development of *Mathematica* to output graphic data in a form suitable for the graphic hardware and software present on graphic workstations. The first of these programs is Live.m, distributed with the copies of *Mathematica* that run on such graphic workstations. We shall

11.3 Converting *Mathematica* Graphics

use similar techniques to generate ray-tracing formats for two popular freely available programs, POVRAY [48] and RAYSHADE [33].

Because ray tracing is a slow process, we do not provide for a *MathLink* connection to the ray-tracing program. We simply write a file that can be input by the ray-tracing program. Because these programs allow many more surface specifications than *Mathematica*, the files need to be postprocessed, in general. *Mathematica* delivers the raw data for the objects; anything else is added by hand.

The following discussion explains the general techniques for writing foreign file formats using as an example the file format expected by POVRAY. The code is in the package POVray.m, reproduced fully in Listing 11.6–1 at the end of this chapter. The conversion for RAYSHADE is similar, and is given in the package rayshade.m (not listed). You should be able to write the necessary code for any other program you might want to use. Hopefully, the program in question will have adequate documentation, explaining the input syntax in sufficient detail.

The conversion function POVray is called in this way:

$$\text{POVray}[\textit{"file.pov"}, \textit{graphic}, \textit{options}\ldots].$$

The procedure writes a POVRAY input file *file*.pov. The graphic can be a Graphics3D, SurfaceGraphics3D, or a SurfaceGraphics (as output by Plot3D). First, some global aspects of the picture are determined, such as light sources, and the plot range. Then, the conversion function maps itself onto the list of graphics primitives present in Graphics3D or it converts the given list of vertices and normals (for SurfaceGraphics3D). A SurfaceGraphics is simply converted to Graphics3D (this conversion is built in) and then treated as such. This driver code for Graphics3D is shown in Listing 11.3–1.

```
writeobj[file_, g:Graphics3D[oblist_, opts_List], myopts___] :=
    Module[{pr, vc, vv, extra},
        newline[file];
        extra = ExtraInformation /. {myopts};
        If[ extra,
            pr = FullOptions[g, PlotRange];
            vc = FullOptions[g, ViewCenter];
            vv = FullOptions[g, ViewVertical];
            processOpts[file, pr, ViewCenter->vc, ViewVertical->vv, opts];
        ];
        newline[file];
        Write[file, "/* begin Graphics3D */"];
        writeobj[file, Chop[N[oblist]], myopts];
        Write[file, "/* end Graphics3D */"]
    ]
```

Listing 11.3–1: Conversion of Graphics3D for POVRAY

If the option ExtraInformation is given, the code uses FullOptions to determine the values of PlotRange, ViewCenter, and ViewVertical that *Mathematica* would use to display the graphic. The auxiliary procedure `processOpts` then generates the declarations for the camera and the light sources, by first converting the special "box coordinates" used by *Mathematica* into world coordinates and then writing out the information. Please refer to Listing 11.6–1 for the details.

The auxiliary function `writeVector[`*stream, text, vector*`]` writes out a vector of three components. It is the main output routine, because almost all data is made of such vectors. Numbers are written out in `FortranForm`, with 5 digits.

In general, you will find that the default light sources that work well for *Mathematica*'s own graphic rendering do not work well for ray tracing. In most cases, you will have to define your own lights.

Graphics primitives are output with the procedure

$$\texttt{writeobj[}\textit{stream, object, options}\texttt{]} \ .$$

It is listable and, therefore, can map itself onto any nested list of graphic objects encountered. The following graphic primitives are converted:

- Polygons are subdivided into triangles. Each triangle is then turned into the POVRAY primitive `triangle`.

- Cuboids become `box` objects.

- A `SurfaceGraphics3D` with normal vectors becomes a number of `smooth_triangle` objects.

- A `SurfaceGraphics3D` without normal vectors becomes a number of `triangle` objects.

- A `SurfaceGraphics` is converted into `Graphics3D` and then processed again.

- Other items on the graphic list are ignored.

Polygons need to be divided into triangles, because POVRAY handles only triangles, rather than arbitrary polygons. Input triangles need no subdivision, quadrilaterals are converted into two triangles, and the general n-gon is converted into n triangles, their center of gravity being the additional common vertex.

We read in the conversion functions for the program POVRAY.	`In[6]:= Needs["MathProg`POVray`"]`
This command was used to prepare the data for Plate 7-c. The output file ray3.pov was then postprocessed by hand to add the additional elements present in the scene of Plate 7-d.	`In[7]:= POVray["/tmp/ray3.pov", %4, ExtraInformation->False]` `Out[7]= /tmp/ray3.pov`

11.4 Sample Images

The color plates give an overview of the effects that can be obtained with ray tracing. Plates 9 through 11-b were rendered with POVRAY. Plate 12 was rendered with RAYSHADE. POVRAY produces its output in the TARGA image file format; RAYSHADE outputs RLE files. These files were converted to ppm format and, finally, to TIFF.

11.4.1 The *Mathematica* Icon

Plate 9 shows the familiar *Mathematica* icon, here made from solid wood floating in the ocean. Note the intricate wave patterns, which can be generated easily with fractal functions built into POVRAY (see also Chapter 8). The clouds in the sky are the result of a partially transparent color map applied to a random spatial function. The image has a resolution of 2250 × 2775 pixels. It was rendered in 3 hours on a SPARCstation 5 and then converted into a 10-MB TIFF file. This image also appeared on the cover of *The Mathematica Journal*, Vol. 4, No. 3.

The raw data were prepared with the polyhedra package.

    In[1]:= Needs["Graphics`Polyhedra`"]

The icon was wrongly called the "stellated icosahedron." It should properly be called a *faceted icosahedron* (see Section 10.4).

    In[2]:= Show[Stellate[Polyhedron[Icosahedron]]];

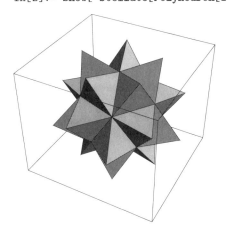

11.4.2 A Soccer Ball

The truncated icosahedron (or soccer ball), one of the uniform polyhedra (see Chapter 9), consists of 12 pentagons and 20 hexagons. Plate 10 shows it with lead-crystal material properties, that is, highly transparent with a high index of refraction. The image shows multiple refractions. You can also see the effects of total internal reflection.

We use the package from Chapter 9.

`In[1]:= Needs["MathProg`UniformPolyhedra`"]`

This figure provided the data for ray tracing.

`In[2]:= Show[ Graphics3D[MakeUniform[w2[2, 5, 3]]] ];`

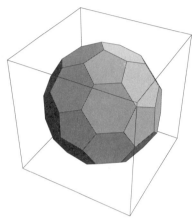

11.4.3 A Minimal Surface

Plate 11-a is one of the minimal surfaces from Chapter 9 of the first volume, rendered with a highly polished brass texture, above a wooden floor.

This package from the first volume also is included on the *Mathematica Programmer* CD-ROM.

`In[1]:= Needs["MathProg`MinimalSurfaces`"]`

The small positive value for r avoids the logarithmic singularity at 0. There is one curl for each 2π increment of the φ range.

```
In[2]:= PolarSurface[z, 1/z, z, {0.0000001, 2}, {-4Pi, 4Pi},
            ViewPoint -> {-2.1, -1.1, 1.2}, Axes -> None,
            PlotRange -> All, PlotPoints -> {14, 120} ];
```

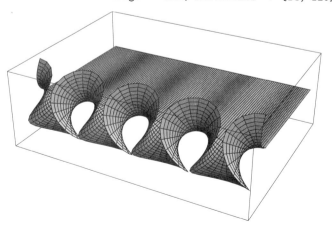

11.4 Sample Images

11.4.4 A Twisted Tube

Plate 11-b shows a torus knot, one of C. Henry Edwards' twisted tubes [18], with a wooden texture. (Plate 14-b shows the same tube as a single-image stereogram, see Section 12.5.2.)

We load the package from the electronic supplement of *The Mathematica Journal*.

```
In[1]:= << TMJ-3.1/Graphics_Gallery/Edwards/TwistedTubes.m
```

Here is the parametric representation for an off-center circle.

```
In[2]:= With[{r5 = 7/10},
            x5[t_] := 1 + r5 Cos[t];
            y5[t_] := r5 Sin[t];
        ]
```

The circle is rotated around the z-axis in a distance 3. There is a twist of 3/2 per revolution and we perform two revolutions.

```
In[3]:= tube[ {x5, y5}, {3, 3/2, 2},
            PlotPoints->{16, 100} ];
```

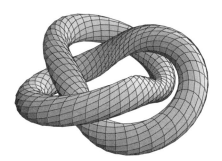

11.4.5 A Shell

The shell in Plate 12 was inspired by a similar graphic by S. Dickson in the second edition of the *Mathematica* manual [60].

We need the parametric plot extension from the standard package.

```
In[1]:= Needs["Graphics`ParametricPlot3D`"]
```

Here are the parameters of the shell.

```
In[2]:= r8[t_] := 1/100 + 1/4 t/(2Pi);\
        z8[t_] := -9/32 (t/(2Pi))^2;\
        rev = 3 + 5/8;\
        r0 = 1/20;
```

The various "fudge factors" prevent degenerate triangles and inconsistent normal vectors.

```
In[3]:= ParametricPlot3D[
          {(r0 + r8[t](105/100 + Cos[phi]))Cos[t],
           r8[t](r0 + r8[t](105/100 + Cos[phi]))Sin[t],
           z8[t] + r8[t]Sin[phi]},
          {t, 0.1, 2Pi rev, 2Pi/22},
          {phi, -Pi+0.1, Pi+0.1, 2Pi/24},
          Axes -> None, PlotRange -> All,
          ViewPoint -> {0, -3, 0.1} ];
```

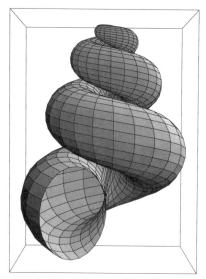

The surface of the shell shows soft specular highlights from the three light sources present. Some of the light sources are extended, that is, they are not (unrealistic) point lights, but have a finite diameter (like a real light bulb). As a consequence, the shadows are soft, too. This effect is achieved by *supersampling*, where several rays in slightly different directions are used to find the final color value of a pixel. The background is a half cylinder. These elements were added by hand to the data generated with Rayshade[].

11.4.6 Sierpinski Sponge

The last image, Plate 16, shows a variant of the Sierpinski sponge. The ordinary Sierpinski sponge is constructed by subdividing a cube into 27 cubes and removing the cubes in the center of the faces and the one in the center of the larger cube. The remaining 20 cubes are then subdivided in the same way. In our variant, we replaced the removed cubes by spheres. The largest spheres at the first step are given a highly refractive green glass texture, the spheres at the second step have a highly reflective silver finish, the spheres at the third step are red with a shiny surface, and, finally, the smallest spheres are blue and show dull highlights. The scene contains 58,947 spheres. Rendering it at a resolution of 2250 × 2775 pixels took several days of CPU time.

11.5 Stereo Pairs

A stereo pair is a set of two pictures of the same object, taken from slightly different camera positions. If the images are viewed in such a way that each eye sees one of the images, respectively, a true three-dimensional effect is achieved. If the width of the images is less than is the distance of your eyes, you can view the images without any equipment. Let us first produce a stereo pair in *Mathematica*, then use the program RAYSHADE to produce ray-traced stereo pairs.

11.5.1 Stereo Pairs in *Mathematica*

To generate a stereo pair, produce a three-dimensional graphic object, then use Show[] twice, with slightly different viewpoints. Finally you can assemble the two images in a graphics array. For an example, we use the package SphereWalk.m, mentioned on page 4.

Here is a spherical random walk of length 3000.

```
In[1]:= walk = SphereWalk[ 3000, 0.11 ]
Out[1]= -Graphics-
```

This is the desired eye distance in *Mathematica*'s box coordinates.

```
In[2]:= ed = 0.2;
```

Here are the necessary options for rendering the two images.

```
In[3]:= opts = {DisplayFunction -> Identity,
            Boxed -> False, SphericalRegion -> True};
```

We generate two images with different viewpoints and view centers (to ensure a parallel projection).

```
In[4]:= {Show[ walk, opts, ViewPoint -> {-ed/2, 2.3, 2},
            ViewCenter -> 0.5 + {-ed/2,0,0} ],
       Show[ walk, opts, ViewPoint -> { ed/2, 2.3, 2},
            ViewCenter -> 0.5 + { ed/2,0,0} ] };
```

We combine the two graphics side by side. The negative value of the GraphicsSpacing option of GraphicsArray brings the two images closer together. View the images by defocusing your eyes until you begin to see four images. Then bring the middle two images together (concentrate on the central dot).

```
In[5]:= Show[ GraphicsArray[%, GraphicsSpacing -> -0.4 ];
```

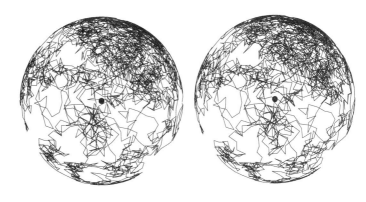

11.5.2 Rayshade Images

The program RAYSHADE provides for an easy way to generate ray-traced stereo pairs. The parameter -E *sep* gives the desired separation of the two camera positions, and the options -l and -r generate the left and right views, respectively. We discuss three examples of stereo pairs rendered with RAYSHADE.

The first example on Plate 4-a uses the great icosidodecahedron (see Chapter 9, Section 12.5.1, and the cover of the first volume for additional renderings of this polyhedron). The graphic data were converted into the RAYSHADE input file poly.ray using the package rayshade.m in much the same way as in Section 12.5.1. The two images were then rendered with the commands

```
rayshade -a -S 4 -R 600 600 -E 0.3 -l -O poly-l.rle poly.ray
rayshade -a -S 4 -R 600 600 -E 0.3 -r -O poly-r.rle poly.ray
```

The resulting files poly-l.rle and poly-r.rle were then converted into TIFF form for inclusion in the color plates.

Plate 4-b shows one of the stellated icosahedra (see Chapter 10). It is no. 19 on the list of icosahedra sorted by size.

Last, Plate 4-c shows a self-avoiding random walk in three dimensions. At each step, An unoccupied site among the six orthogonally adjacent sites to the current site is chosen at random, with sites closer to the origin having a higher probability than sites farther away. The path is then converted into a cylindrical tube. The result is a rather compact plumbing structure. The code to produce the tubes is in the file RayWalk.m. Methods to construct random walks are discussed in Section 1.3.2.

11.5.3 Wallpaper Stereograms

Another visualization technique uses several images, aligned horizontally. The viewpoint rotates by a constant angle from one image to the next. When you focus your eyes such that the eyes look at adjacent images, you will see a row of three-dimensional objects. Plate 13 shows three uniform polyhedra (nos. 33, 54, and 60, see Chapter 9) with different spacing and rotation angles, giving the effect of different distances from the observer. The solids have been rendered with POVRAY. The background is a wallpaper stereogram, too. It is a repeating pattern, which gives the impression of a flat surface in some distance behind the solids. In fact, the background is an image of the floor in ETH's mechanical engineering building!

The CD-ROM contains the packages POVray.m, rayshade.m, RayWalk.m, and SurfaceGraphics3D.m, as well as the files raypic1.pov and raypic8.ray with the POVRAY and RAYSHADE input for Plates 10 and 12, respectively.

11.6 The Complete Code of POVray.m

```
BeginPackage["MathProg`POVray`"]

POVray::usage = "POVray[filename, graphics] writes graphics in POVray format
        to file. POVray[stream, graphics] uses stream for output."
POVHeader::usage = "POVHeader->string" is an option of POVray that gives
        the string to write at the beginning of the output file."
POVInformation::usage = "POVInformation->Automatic is an option of POVray that specifies
        which extra information is to be included. By default, the light sources and
        camera position are output. POVInformation->None prints no extra information."

Options[POVray] = {
        POVHeader -> "#default { texture { pigment {color rgb <1,1,1>}
                                finish {phong 0.8 ambient 0.4} }}",
        POVInformation -> Automatic
}

Begin["`Private`"]

Needs["Utilities`FilterOptions`"]

POVray[filename_String, obj_, opts___] :=
    Module[{file},
        file = OpenWrite[filename, FormatType -> OutputForm, PageWidth -> Infinity];
        If[ file === $Failed, Return[file] ];
        POVray[file, obj, opts];
        Close[file]
    ]

POVray[file_OutputStream, obj_, opts___] :=
    Module[{prolog, extra},
        prolog = POVHeader /. {opts} /. Options[POVray];
        If[ prolog =!= "", Write[file, prolog] ];
        extra = POVInformation /. {opts} /. Options[POVray];
        If[ extra === Automatic, extra = True ];
        writeobj[file, obj, POVInformation -> extra, opts];
        file
    ]

newline[file_] := Write[file, ""]

(* fix Infix problem for functions with one argument *)
protected = Unprotect[Infix]
Infix[_[e_], h_:Null] := e
Protect[ Evaluate[protected] ]

(* optional text followed by numbers, FortranForm with 5 digits *)
ndig = 5
nForm[r_] := NumberForm[FortranForm[r], ndig]
writeNums[file_, txt_:"", r_?NumberQ] := Write[file, txt, " ", nForm[r]]
writeNums[file_, txt_:"", r_List] := Write[file, txt, " ", Infix[nForm /@ r, " "]]
```

```
(* POV-ray vector notation *)
writeVector[file_, txt_:"", r_List] /; Length[r] == 3 :=
    Write[file, txt, "<", Infix[nForm /@ r, ", "], ">"]
(* write various graphic objects *)
SetAttributes[writeobj, Listable]
(* Graphics3D *)
writeobj[file_, g:Graphics3D[oblist_, opts_List:{}], myopts___] :=
    Module[{pr, vc, vv},
        newline[file];
        If[ POVInformation /. {myopts},
            pr = FullOptions[g, PlotRange];
            vc = FullOptions[g, ViewCenter];
            vv = FullOptions[g, ViewVertical];
            processOpts[file, pr, ViewCenter->vc, ViewVertical->vv, opts];
        ];
        newline[file];
        Write[file, "/* begin Graphics3D */"];
        writeobj[file, Chop[N[oblist]], myopts];
        Write[file, "/* end Graphics3D */"]
    ]
writeobj[file_, Polygon[vl_], myopts___] := writeTriangle[file, #]& /@ toTriangles[vl]
writeobj[file_, Cuboid[p1_], myopts___] := writeobj[file, Cuboid[p1, p1 + {1,1,1}]]
writeobj[file_, Cuboid[p1_, p2_], myopts___] := (
        newline[file];
        Write[file, "box {"];
        writeVector[file, "\t", p1];
        writeVector[file, "\t", p2];
        Write[file, "}"];
    )
(* SurfaceGraphics3D from SurfaceGraphics3D.m *)
writeobj[file_, g0:MathProg`SurfaceGraphics3D`SurfaceGraphics3D[vl_, norms_, opts_],
        myopts___] :=
    Module[{pr, vc, vv, g},
        newline[file];
        If[ extra = POVInformation /. {myopts},
            g = Graphics3D[g0];
            pr = FullOptions[g, PlotRange];
            vc = FullOptions[g, ViewCenter];
            vv = FullOptions[g, ViewVertical];
            processOpts[file, pr, ViewCenter->vc, ViewVertical->vv, opts];
        ];
        newline[file];
        Write[file, "/* begin SurfaceGraphics3D */"];
        Scan[ writeSmoothTriangles[file, #[[1]], #[[2]]]&,
              Chop[Transpose[{makeMesh[vl], makeMesh[norms]}]] ];
        Write[file, "/* end SurfaceGraphics3D */"]
```

```
    ]
(* without normals: turn into Graphics3D without loss of information *)
writeobj[file_, g0:_MathProg`SurfaceGraphics3D`SurfaceGraphics3D, myopts___] :=
    writeobj[file, Graphics3D[g0], myopts]
(* SurfaceGraphics: turn into Graphics3D without loss of information *)
writeobj[file_, g0:_SurfaceGraphics, myopts___] := writeobj[file, Graphics3D[g0], myopts]
(* add other graphics types/primitives here *)
(* catch all *)
writeobj[file_, _, myopts___] := Null (* unknown object *)
makeMesh[vl_List] :=
    Module[{l = Drop[#, -1]& /@ vl, l1 = Drop[#, 1]& /@ vl, mesh},
        mesh = {Drop[l, -1], Drop[l1, -1], Drop[l1, 1], Drop[l, 1]};
        Transpose[ Flatten[#, 1]& /@ mesh ]
    ]
(* turn n-gon into list of triangles *)
toTriangles[vlist_List] /; Length[vlist] == 3 := {vlist}
(* treat 4gons specially for efficiency *)
toTriangles[vlist_List] /; Length[vlist] == 4 :=
        {vlist[[{1, 2, 3}]], vlist[[{3, 4, 1}]]}
(* general case: use center of gravity as additional vertex *)
(* in this way, some nonconvex polygons also can be rendered correctly *)
toTriangles[vlist_List] :=
    Module[{bary = (Plus @@ vlist)/Length[vlist], circ},
        circ = Partition[ Append[vlist, First[vlist]], 2, 1 ];
        Apply[ {#1, #2, bary}&, circ, {1} ]
    ]
writeSmoothTriangles[file_, {a_, b_, c_, d_}, {na_, nb_, nc_, nd_}] := (
        writeSmoothTriangle[file, d, nd, c, nc, a, na];
        writeSmoothTriangle[file, b, nb, a, na, c, nc];
    )
writeTriangle[file_, vl_List] := (
    Write[file, "triangle {"];
    writeVector[file, "\t", #]& /@ vl;
    Write[file, "}"]
)
writeSmoothTriangle[file_, p1_, n1_, p2_, n2_, p3_, n3_] := (
        Write[file, "smooth_triangle {"];
        writeVector[file, "\t ", p1];
        Scan[ writeVector[file, "\t, ", #]&, {n1, p2, n2, p3, n3} ];
        Write[file, "}"];
    )
```

```
(* box to world coordinates *)
scale[{min_, max_}, rat_] := (min+max)/2 + (max-min)/2 rat
boxScale[pr_, pos_] := MapThread[ scale, {pr, pos} ]

cam = 3.0 (* approximate field of view *)

processOpts[file_, pr_, opts___] :=
    Module[{optlist = Flatten[{opts}], prlow, prhigh,
            vp, vc, vv, svp, svc, svv, fow, lights},
        Write[file, "/* plot range:"];
        writeNums[file, " * ", #]& /@ pr;
        Write[file, " */"];
        {prlow, prhigh} = Transpose[pr];
        vp = ViewPoint /. optlist /. Options[Graphics3D];
        svp = boxScale[ pr, vp ];
        vc = ViewCenter /. optlist /. Options[Graphics3D];
        svc = prlow + vc (prhigh - prlow);
        vv = ViewVertical /. optlist /. Options[Graphics3D];
        svv = boxScale[ pr, vv ];
        fow = Sqrt[Plus @@ (vp^2)]; (* focal length *)
        Write[file, "camera {"];
        writeVector[file, "\t location ", svp];
        writeVector[file, "\t direction ", {0, fow, 0}];
        writeVector[file, "\t up ", {0,0,cam}];
        writeVector[file, "\t right ", {cam,0,0}];
        writeVector[file, "\t sky ", vv];
        writeVector[file, "\t look_at ", svc];
        Write[file, "}"];
        lights = LightSources /. optlist /. Options[Graphics3D];
        Scan[doLight[file, pr, #]&, lights];
    ]

doLight[file_, pr_, {pos_, RGBColor[r_, g_, b_]}] :=
    Module[{scaled},
        scaled = boxScale[ pr, pos ];
        Write[file, "light_source { " ];
        writeVector[file, "\t", scaled];
        writeVector[file, "\t color rgb ", {r, g, b} ];
        Write[file, "}"];
    ]
End[]

Protect[Evaluate[$Context <> "*"]]

EndPackage[]
```

Listing 11.6–1: POVray.m

Chapter Twelve

Single-Image Stereograms

It has been known for over 20 years that single images can be constructed that allow true three-dimensional viewing. The technique of the single-image stereogram (SIS) has gained much attention in the last few years, due to the amazing success of a series of books with such images. The basic construction of stereograms is simple and can easily be programmed in *Mathematica*. Alternatively, graphics objects can be written out in a form suitable as input for external stereogram generators.

Section 1 gives an introduction to stereograms. In the next section we develop a program to generate random-dot stereograms in *Mathematica*. A discussion on the design of good stereograms follows in Section 3. Next, in Section 4, we show a technique for producing much more accurate stereograms, using some of *Mathematica*'s root-finding capabilities. In Section 5 we develop code to convert graphics into input for an external stereogram generator, using techniques developed originally for ray tracing in Chapter 11.

A fundamental limitation of the SIS technique is the lack of surface texture. It is impossible to give the surfaces in the scene their own color or shading. Nevertheless, the ability to view true three-dimensional images without any special equipment, images that can be displayed on a computer screen and stored electronically, is interesting and could be used for teaching and the visualization of scientific data. It is more than a toy.

About the illustration overleaf:
A stereogram of the function $\frac{xy^2}{x^2+y^4}$. This function has limit 0 for $(x, y) \to (0, 0)$, independent of the direction of approach, but it is not differentiable at $(0, 0)$.

The code for this picture is in BookPictures.m.

12.1 Introduction

Single-image stereograms (SIS) have become quite popular, mainly due to the success of the books in the "Magic Eye" series [43, etc.]. *The Mathematica Journal* published an article about single-image random-dot stereograms (SIRDS) in 1991 [4]. A good overview of the history of stereograms is given in the book [30]. It shows a larger variety of stereograms and related techniques than the Magic Eye books. In its chapter on the history of SIS, you can see the very first random-dot stereograms ever produced. The first stereograms consisted of two images, comparable to stereo pairs of photographs, which have been around for over a century. The book contains some examples of these stereo pairs, as well as some stereograms of artistic merit. Unavoidably, this book, too, has spawned a sequel, *Super Stereograms*.

The breakthrough was the development of the *single-image* stereogram. In the next section, we shall show how the simplest of these, SIRDS, is generated from an array of depth data. The placement of the random dots into a rigid matrix introduces roundoff errors that limit the depth resolution available. Other methods are needed to overcome these limitations.

If we are given a function of two arguments that describes the depth at any given point, instead of a matrix of depth values at certain points only, *Mathematica* allows us to compute depth data to arbitrary numerical precision. Using numerical root finding, we can compute exact placements for graphical objects that make up a SIS. We develop a versatile program for this kind of SIS and show some of the effects possible.

Finally, we look at one external SIS generator, RAYSIS, a freely available program that produces high-quality images. Because the syntax of its scene description is similar to the language of the ray-tracing program POVRAY, described in Chapter 11, it was easy to write a *Mathematica*-to-RAYSIS conversion program. Along the way, we discuss a few useful programming ideas.

12.2 The Classic SIRDS in *Mathematica*

The viewing model for a standard SIRDS assumes a scene, given by a rectangular array of depth or z values. You can think of such a scene as the graph of a function of two variables, viewed from straight above (that is, the viewpoint lies on the positive z-axis). Between the eye and the scene is the image plane, where the stereogram will be formed. A typical cross section in the x-z plane of such a setup is shown in Figure 12.2–1. This kind of stereogram for which you focus your eyes behind the image plane is called a *wall-eye stereogram*. Less popular and more difficult to view is the second kind, a *cross-eye stereogram*, for which you focus in front of the image plane.

For practical reasons, the depth data is confined between the *far plane* (at $z = 0$) and the plane halfway between the far plane and the image plane. The image plane is assumed to be at $z = d_1$, a distance d_2 from the eyes. An important parameter is the *eye separation*, denoted by e. Each

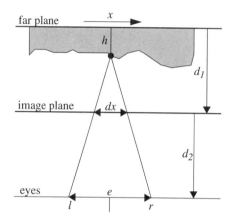

Figure 12.2–1: Determining dot distance

point p in the scene is viewed by both eyes, and the rays from the eyes to p intersect the image plane in two locations, separated by a distance dx (the two intersection points have the same y value). It is easy to compute dx from the height h of p and the parameters d_1, d_2, and e:

$$dx = e \frac{d_1 - h}{d_1 + d_2 - h}. \qquad (12.2\text{--}1)$$

Assuming we have a rectangular array of height values h_{ij}, with $1 \leq i \leq n_y$ and $1 \leq j \leq n_x$, we can perform this calculation for each h_{ij} to obtain a matrix of distances, dx_{ij}.

Now, we want to generate a raster image with the same dimensions, $n_y \times n_x$, which can be viewed in such a way that the three-dimensional scene becomes visible. A necessary condition is that the pairs of points $(j \pm dx_{ij}/2, i)$ have the same color or grayscale value, because they correspond to one point in the scene, seen with both eyes. The values $j \pm dx_{ij}/2$ must be rounded to the nearest integer, to stay within the raster. Dots not constrained by this condition should have uncorrelated values. The fact that this condition is also sufficient, that is, it allows our brain to see a three-dimensional image if the eyes are properly focused behind the image plane, is what makes SIRDS so amazing.

Enforcing the restrictions on the pixel values can be done separately for each i, that is, for each horizontal scan line. Several methods can be used. One way is to compute, for each j, $1 \leq j \leq n_x$, the pair of values $\{k, r(k)\}$ where $k = j - d_{ij}/2$ and $r(k) = j + d_{ij}/2$. Then assign pixel values v_j for $j = n_x, n_x - 1, \ldots, 1$ by setting $v_j = v_{r(j)}$ if $j = k$ for some pair $\{k, r(k)\}$ with $1 \leq r(k) \leq n_x$, and choosing a random value for v_j otherwise. This computation must be performed sequentially, from $j = n_x$ down to $j = 1$, that is, from right to left. For a SIRDS, the random values are black or white dots, chosen with equal probability.

Matrices of depth information are readily available in *Mathematica*. The data structures of `SurfaceGraphics`, `DensityGraphics`, and `ContourGraphics` all contain such a matrix. Equation 12.2–1, expressed as a function of the height h, can simply be mapped to all elements of

12.2 The Classic SIRDS in *Mathematica*

the matrix. Then, for each line (fixed i), we build the set of rules

$$j - dx_j/2 \ \text{->} \ j + dx_j/2 \qquad (12.2\text{--}2)$$

for $j = 1, 2, \ldots, n_x$. The value $r(j)$ is then found by applying these rules to j. If the result is not in the range $1 \ldots n_x$ or is equal to j, a random value is used. The code of SIRDS[*graphic*, *options*] is in the package SIS.m, reproduced in Listing 12.6–1 at the end of this chapter. Here is a first example of its use.

This command generates a suitable -SurfaceGraphics- object. Note that the settings for PlotPoints should be large and in the same proportion as the plot ranges in the x and y directions. For reasons of efficiency, this draft plot has been generated with a smaller number of plot points.

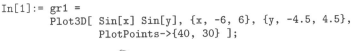

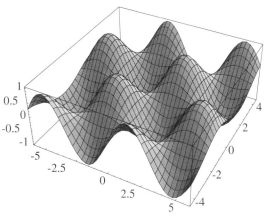

Figure 12.2–2 shows the resulting SIRDS. Graphic gr1a is the same as gr1, but with a higher resolution. PlotPoints -> {200, 150} was used to generate it.

In[2]:= SIRDS[gr1a];

The two guide dots at the top may help you see it. Good hints for viewing stereograms are given in most books containing such images. I find it easiest to casually look at the image, then let my eyes relax, causing four images of the two guide dots to appear, and then slowly change focus until two of the four images meet in the middle. Then, I shift my attention downward, without trying to refocus my eyes. The important point is not to try to change your gaze once the image starts to appear. Take your time. With some practice, you will no longer need the guide dots.

This method generates rather crude images. If the eye separation is E pixels and $d_1 = d_2$, the minimum and maximum values of dx (in numbers of pixels) are $dx_{min} = E/3$ and $dx_{max} = E/2$, for $h = d_1/2$ and $h = 0$, respectively. Figure 12.2–2 has $n_x = 200$ pixels per line and an eye distance of 70, resulting in only 12 discrete depth levels ($dx_{max} - dx_{min}$). However, at print resolution (200 dots per inch, say), satisfactory images can be obtained.

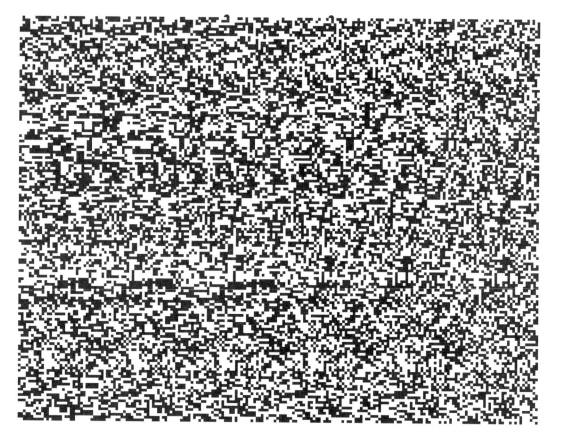

Figure 12.2–2: $\sin x \sin y$

12.2 The Classic SIRDS in *Mathematica*

A density graphic can be interpreted as a grayscale coding of height values. Many commercial and freely available SIRDS generators take their input in the form of such grayscale images. This function generates a cone, viewed from above. Figure 12.2–3 shows the cone as a SIRDS with a horizontal resolution of 600 dots. The computation time on a SPARCstation 5 was 10 minutes for the density plot and 14 minutes for the SIRDS.

```
In[3]:= pi2 = N[2Pi]; \
    cone =
    DensityPlot[ 7 - Sqrt[x^2+y^2],
        {x, -6, 6}, {y, -4, 4},
        PlotPoints -> 25{6, 4},
        AspectRatio -> Automatic, Mesh -> False
    ];
```

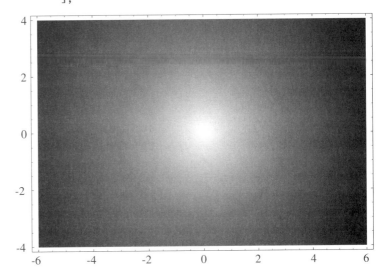

A few notes on the code of the command SIRDS. The command accepts a graphic object of type SurfaceGraphics, ContourGraphics, or DensityGraphics. The internal structure of these three graphic types is essentially the same. We are interested only in the matrix of height values. The whole graphic object is named (gr), so we can use FullOptions to obtain the values for the plot range that *Mathematica* uses to display the object. In hmax we store the range of height values and scale the other quantities accordingly.

The function SIRDSLine computes dots for one line; it is simply mapped onto the rows of the matrix zmat. Note that Equation 12.2–1 is copied almost verbatim. The variable line is a list, and calculations are performed for each point without the use of a clumsy loop. The expression Range[n]-0.5dx builds the left sides of the rules from Formula 12.2–2. The right sides are simply left + dx, properly rounded. Thread builds the list of rules. The rules will be used many times, and they have constant left sides (that is, they do not contain pattern variables); therefore, using Dispatch speeds up the computations a bit. The loop over j is necessary, because we need sequential evaluation. SIRDSLine is declared locally within SIRDS because it needs access to many precomputed values. Had it been declared outside the body of SIRDS, all these values would have to be passed as parameters.

The graphic primitive Raster[*mat*] takes a matrix of gray values (between 0 and 1). Therefore, a random dot is simply either 0 or 1, with equal probability, that is, Random[Integer].

Figure 12.2–3: A cone

12.3 Designing Good Images

To aid in viewing a stereogram, the range of the height values should not be too large in relation to the distance to the image plane. A maximum value of $d_1/2$ is the largest one you should try. With the recommended setup of $d_1 = d_2$, $h_{max} = d_1/2$, the far plane appears to be at twice the distance from the eyes to the image plane, where the stereogram is rendered. The three-dimensional image will occupy the space from the far plane to halfway between the far plane and the image plane.

SIRDS uses the plot range of the given graphic object to find the minimum and maximum values of h that appear, and clips values outside this range. The values d_1 and d_2 can be changed by setting the options PlaneDistance and EyeDistance, respectively. The values are given in scaled coordinates, where h_{max} is the unit. The defaults $d_1 = 2$ and $d_2 = 2$ correspond to the recommendations about the image geometry given in the preceding paragraph. Often, this range of height values is still too large. You can use PlaneDistance -> 4, EyeDistance -> 4, for

12.3 Designing Good Images

Figure 12.3–1: $\sin x \sin y$ with less apparent depth

example, to halve the apparent depth range. Figure 12.3–1 shows the resulting SIRDS, whose depth appears reduced compared with Figure 12.2–2. The graphic used for Figure 12.3–1 is the same as `gr1` and `gr1a`, but with an even higher resolution. `PlotPoints -> {800, 600}` was used to generate it.

If the object is simply the far plane itself, that is, all $h_{ij} = 0$, the distances dx_{ij} are all equal to $e/2$, half the distance of the eyes. The effect is a repeating random pattern of width $E/2$ (remember that E is the eye distance, expressed in pixels). A consequence of these considerations is that a SIRDS is not scale invariant! It must be generated for a particular desired final size. The images in this chapter are designed for a horizontal dimension of 5.9 inches. The separation of my eyes measures about 2.4 inches, which is a fraction of $s = 0.4$ times the horizontal size. The option `EyeSeparation -> s` can be used to specify the desired separation as a fraction of horizontal size.

If you want to generate your own images and look at them on the screen, you should determine the right value of the eye separation. Under the notebook frontend, you should set the default size of graphics cells so that they are sufficiently wide (at least 5 inches is recommended). Measure the width on the screen and divide your eye separation by the image size. With X11 and its derivatives, you can set the variables `$DisplayWidth` and `$DisplayHeight` to sufficient values, generate a graphic and measure its size. To check whether the setup is right, measure the distance between the guide dots; it should be half your eye separation. Once you have determined the correct value, you can use

$$\text{SetOptions[SIRDS, EyeSeparation -> } s\text{]}$$

to cause all future images to use this value.

If the eye separation is too large, depth information will be magnified and the image may be difficult to see. If the distances dx exceed the eye separation, no three-dimensional image can be seen, because few people can turn their eyes outward. If the eye separation is too small, there is a danger that you will not focus on adjacent equivalent dots, but on every other one. You will still see a three-dimensional effect, but the object will be distorted.

12.4 Exact Stereograms

The method for producing SIRDS in the preceding section suffers from two approximations. First, distances between dots on the image plane belonging to the same point of the scene are rounded to the underlying raster. Second, we used parallel projection, that is, the eyes were always assumed directly above each point. In reality, our eyes stay fixed over the center of the image. Let us try to develop a method that overcomes both limitations.

Bar-Natan [4] described a way to avoid rounding to a raster: Start with one dot at $x = x_0$ near the left margin. Using parallel projection, compute the distance dx_1 appropriate for this dot. Draw a second dot at $x_1 = x_0 + dx_1$. Repeat this step until you exceed the right margin. Draw additional series of dots with starting values in the range $0 < x_0 < e/2$. Perform this operation for several, equidistant values of y. Instead of drawing dots, you can draw any (small) graphic object, such as a black square, centered at the required coordinates. The resulting images no longer show discrete steps in the apparent height values (assuming your output device has a high enough resolution). The asymmetric nature of the procedure can lead to a clustering of dots near the right margin (clearly visible in the image on page 72 of Bar-Natan's article).

We shall use a different approach: We choose each initial dot randomly over the whole image plane. Then we extend the dots to the left and right with the method just described. Further, we trace the rays from a fixed position of the eyes. For this purpose, an array of height values is no longer sufficient. We need the height as a function of x and y. We can use numerical root finding to compute the intersection between the rays and the surface defined by $z = h(x, y)$ to the desired accuracy. The geometry becomes quite a bit more complicated (see Figure 12.4–1). If the point $P_0 = (x_0, y_0)$ has already been constructed, we want to find P_1. We introduce a local coordinate

12.4 Exact Stereograms

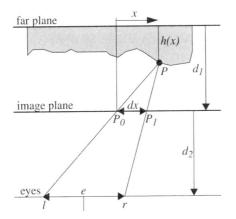

Figure 12.4–1: Skew projection

system with variable x, measured from x_0. The coordinates of a point P on the ray from the left eye through P_0 can be expressed in terms of x as $(x + x_0, y(x), z(x))$, where $y(x)$ and $z(x)$ are simple rational functions of x. If this point lies on the surface, its z coordinate is $h(x + x_0, y(x))$, so we have to solve the equation

$$z(x) = h(x + x_0, y(x)). \tag{12.4--1}$$

Then, we draw a ray from P to the right eye and intersect it with the image plane to obtain $x_1 = x_0 + dx$.

We implemented these computations in a command

$$\text{SIS}[hxy, \{x, x_{min}, x_{max}\}, \{y, y_{min}, y_{max}\}, \text{options}].$$

The expression hxy is the height in terms of the variables x and y. The code is in SIS.m, reproduced in Listing 12.6–1. It contains a few programming ideas worth mentioning.

The values of the options EyeSeparation, EyeDistance, and PlaneDistance are extracted in the same way that they are in the SIRDS command presented earlier. PlotPoints is interpreted as the number of independent random initial dots to use. Because we are not given a number of computed height values, but only the function hxy to compute them, we do not know the plot range *a priori*. Therefore, it must be given using the option PlotRange -> {*far*, *near*}. The default is {0, 1}. Next, we define the auxiliary function next that computes the next point according to Equation 12.4–1. It is declared locally within SIS, so that it has access to the many parameters needed. Only the parameters that depend on the current point are passed as arguments. The variables xlow and xhigh are used as initial values for the secant method employed by FindRoot. (The fact that FindRoot uses the secant method instead of an analytical approach if two initial values are given is not well documented.) It is important that the plot range given be right. If the values of hxy lie significantly outside the given range, FindRoot may not find a solution.

In the main loop in SIS, we select a random starting point and then try to find its neighbors to the left and right. At each computed point, a copy of the desired graphic object is added to the list of objects. The graphic object must be given as a (pure) function that is applied to the coordinate pair {x, y}. The default is Point. Point is not really a function, but if it is applied to such a coordinate pair, we get Point[{x, y}], which is in the right form and describes a point centered at {x, y}. You can give your own object with an option setting of the form Object -> *obj*. To give you even more flexibility, the value of the Object option is evaluated once for each iteration of the main loop. Thus, you can even construct a (delayed) option value (of the form Object :> *obj*) that causes the objects for different initial points to be different. Examples are given later.

Please note that the range of x and y values is measured at the image plane. With the ordinary setup, the corresponding ranges at the far plane are about twice as large. This fact should be kept in mind when figuring out the range of values that *hxy* assumes. This plot range must be valid far beyond the given x and y range. If necessary, you can clip the values with an expression of the form Max[*low*, Min[*high*, *hxy*]]. Because we use the secant method for root finding, such expressions pose no problem.

An often-used example is the sombrero, obtained from a rotation of $J_0(r)$, where $r = \sqrt{x^2 + y^2}$. It is shown in Figure 12.4–2. The settings for PlaneDistance and EyeDistance lessen the amount of vertical distortion.

```
In[4]:= SIS[ BesselJ[0, Sqrt[x^2+y^2]],
             {x, -8, 8}, {y, -6, 6},
             PlotRange -> {-0.4, 1}, PlotPoints -> 500,
             PlotStyle -> PointSize[0.004],
             PlaneDistance -> 3, EyeDistance -> 3 ];
```

Here is a sample generator for object functions suitable for SIS that produces random characters. The character is chosen randomly among all printable ASCII characters (except the space character, of course).

```
In[5]:= randChar :=
         With[{ch = FromCharacterCode[
                      Random[Integer, {33, 126}]]},
              Text[ch, #]& ]
```

Everytime randChar is evaluated it generates a pure function that places a certain character at position {x, y}.

```
In[6]:= randChar
Out[6]= Text[3, #1] &
```

Note the use of :> (RuleDelayed) instead of -> in the setting for Object to avoid the evaluation of randChar when SIS is called. The resulting image is shown in Figure 12.4–3.

```
In[7]:= SIS[Sin[x + Sin[y]], {x, -4, 4}, {y, -2.5, 2.5},
            PlotRange -> {-1, 1}, PlotPoints -> 500,
            Object :> randChar, PlaneDistance -> 3,
            DefaultFont -> {"Courier", 6} ];
```

Listing 12.4–1 shows some more ideas for object generators. I leave it to you to figure out what they do. With is the easiest way to insert certain values into the body of a pure function.

12.5 Interface to External SIS Generators

The image on page 255 was produced with this command. The function $\frac{xy^2}{x^2+y^4}$ has limit 0 for $(x,y) \to (0,0)$, independent of the direction of approach, but it is not differentiable at $(0,0)$.

```
In[8]:= SIS[ x y^2/(x^2 + y^4), {x, -4, 4}, {y, -3, 3},
        PlotRange -> {-1/2, 1/2}, PlotPoints-> 4000,
        PlotStyle -> PointSize[0.001],
        PlaneDistance -> 2.5, EyeDistance -> 2.5,
        EyeSeparation -> 0.4
     ];
```

12.5 Interface to External SIS Generators

So far, we rendered only functions of x and y, which makes it easy to find the z values at every point. What about general three-dimensional graphics? For an arbitrary three-dimensional graphic object, it is not easy to generate the height values. This task amounts to hidden surface elimination

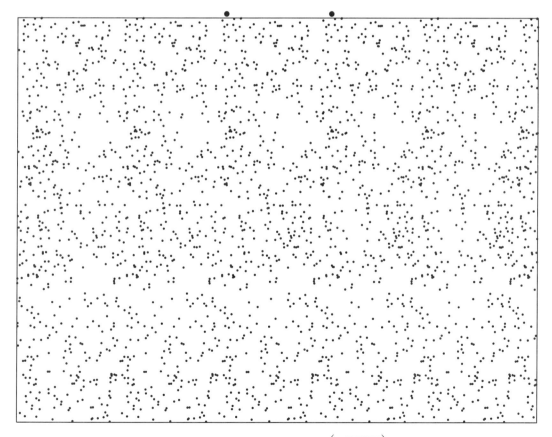

Figure 12.4–2: The sombrero, $J_0\left(\sqrt{x^2+y^2}\right)$

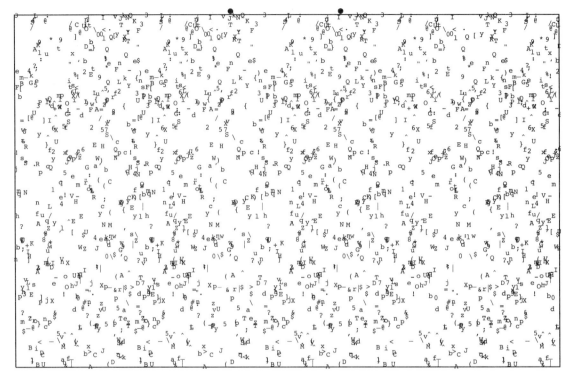

Figure 12.4–3: $\sin(x + \sin y)$

or the computation of a so-called z buffer. Once we have the z values, we can use one of the methods already discussed to render the scene. Most SIS generators take a grayscale image as input and produce a stereogram of some kind. The gray levels encode height: the value 0.0 (black) is assumed to lie at the far plane, the value 1.0 (white) describes the frontmost elements of the scene.

```
randStar :=
  With[{n=Random[Integer, {2, 5}]},
    With[{star = 0.05 Table[Through[{Cos, Sin}[2.0 n Pi i/(2n+1)]], {i, 0, 2n+1}]},
      Function[xy, Line[xy+#& /@ star]]
  ] ]
circ := Function[xy, Circle[xy, 0.04]]
colorCirc := With[{rgb = RGBColor[Random[], Random[], Random[]],
                   rad = 0.04 Random[Real, {0.1, 1}]^1.5},
                  Function[xy, {rgb, Circle[xy, rad]}] ]
randColor := With[{rgb = RGBColor[Random[], Random[], Random[]]},
                  Function[xy, {rgb, Point[xy]}] ]
```

Listing 12.4–1: Object generators

12.5 Interface to External SIS Generators

Some renderers can compute the z values from a three-dimensional scene description. First, they compute a z buffer, then they render the image from these z values. A freely available program in this class is XPGS, which was developed for DOS/Windows and has been ported to UNIX/X11. Its input syntax is rather simple. A list of points is followed by a list of triangles, where the triangles are described by the indices into the list of points for their three vertices. It should not be hard to write a function to convert *Mathematica* polygons into this format.

12.5.1 Stereograms and Ray Tracing

An interesting approach, due to Pascal Massimino, is to use a ray-tracing program to generate the z data. Just as in normal ray tracing, we generate rays emanating from a point in space (the camera in ray tracing; here, the midpoint between the eyes) and find out where they intersect a given three-dimensional scene. Instead of computing shadows, surface colors, and so on, we simply look at the z coordinate of the point of intersection. Once we have this coordinate, we proceed as before in our procedure SIS. Massimino created the program RAYSIS by modifying a popular ray tracer, POVRAY, so that it generates stereograms. Because we developed a package to convert *Mathematica* graphics into POVRAY input in Chapter 11, we are off to an easy start. (Unfortunately, RAYSIS does not use the exact same input syntax as does POVRAY, so some modifications were necessary. This misfeature is likely to be corrected in a future version.)

Our code is in the package RaySis.m. It is similar to the package POVray.m from Chapter 11 and, therefore, not listed here. It provides the command

$$\text{RaySis}[\textit{filename}, \textit{graphic}, \textit{options}]$$

that writes out a three-dimensional graphic scene in RAYSIS input form. The *Mathematica* viewpoint is translated into a RAYSIS camera definition and the far plane is placed behind the bounding box. Note that no clipping at the box surrounding the graphic is performed. I recommend generating all your graphics with the setting `PlotRange -> All` to avoid bad surprises. The field of view and the placement of the far plane may need some manual adjustment; that is, the output file needs to be edited.

Let us show how to create a SIS of a uniform polyhedron, the great icosidodecahedron (my favorite).

We read in the conversion functions.	`In[9]:= Needs["MathProg`RaySis`"]`
Functions for generating all uniform polyhedra are described in Chapter 9. They are defined in this package.	`In[10]:= Needs["MathProg`UniformPolyhedra`"]`

The polyhedron is generated from its Wythoff symbol $2|\frac{5}{2}\ 3$.

`In[11]:= Show[Graphics3D[MakeUniform[w1[2, 5/2, 3]]]];`

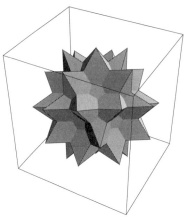

We write out its definition in the form required by RAYSIS.

`In[12]:= RaySis["/tmp/unipoly.sis", %]`

`Out[12]= /tmp/unipoly.sis`

Listing 12.5–1 shows the top of the file unipoly.sis.

```
/* plot range:
 *   -1.6989  1.6989
 *   -1.5196  1.5196
 *   -1.7864  1.7864
 * ViewPoint:   1.3 -2.4 2.
 */
camera {
    /* POSITION */  <4.6446, -8.5746, 7.1455>
    /* LOOK_AT  */  <-1.0013, 1.8486, -1.5405>
    /* UP       */  <0., 0., 3.5727>
    /* lx       */      6.3711
}
/* begin Graphics3D */
triangle {
    <0, 0, 1.7013>
    <0.61803, 0, 0.85065>
    <0.27639, 0.55279, 0.85065>
}
...
```

Listing 12.5–1: Part of unipoly.sis, a RAYSIS input file

The comment at the top echoes some of *Mathematica*'s internal parameters, to help in editing the file. You can modify the field of view by changing the value of `lx` in the camera definition. This prologue is followed by the triangles that constitute the graphic object.

12.5 Interface to External SIS Generators

Because RAYSIS is basically a ray tracer, it takes some time to render an image. Here is the command used to render unipoly.sis:

```
raysis unipoly.sis -W 1770 -H 1200 -P 336 +o unipoly.ppm
```

RAYSIS generates its output as an rgb file in the ppm (portable pixel map, part of PBM) format. The resulting stereogram is shown in Plate 14-a. The most important parameter is, again, the width of the pattern. It should be half the eye distance. This image is printed at 300 dots per inch (giving a horizontal size of 5.9 inches). The eye separation is 2.4 inches. Half of this gives about 336 pixels, the value of the -P option. The -W and -H options are, of course, the width and height of the image. The computation took 20 minutes on a SPARCstation 20. The color image is similar to the SIS found in the many books and posters using similar techniques. RAYSIS uses a clever interpolation method to overcome the limited number of depth levels often found in such images. This kind of stereogram, which modifies an initial seed pattern, is often termed a *wallpaper stereogram*. (A different three-dimensional rendering of the great icosidodecahedron is shown on Plate 4-a; See also Section 11.5.)

12.5.2 Pattern Generation

RAYSIS can generate suitable patterns of any width or it can use an external image (in ppm format). One choice is to use an image with recognizable two-dimensional objects (or even a digitized photo), unrelated to the three-dimensional image hidden in the stereogram. Another approach is to use a random-looking pattern. One source of such random patterns is a generator of fractal images. We used the Fourier synthesis method to generate fractional Brownian motion (fBm) data from Section 8.3 to generate random color patterns. The algorithm generates a matrix of random Fourier coefficients with a distribution chosen to give the desired fractal dimension (between 2 and 3). For generating random landscapes, we would choose the coefficients in such a way that the result of applying the inverse Fourier transform is real valued. Here, we do not impose such a restriction, giving terms with random phase. We then interpret the phase as a color value. The magnitude of the complex numbers is interpreted as brightness, taken modulo 1. Therefore, the formula used is Hue[Arg[z]/(2Pi), 1, Mod[Abs[z], 1.0]]. A pattern is generated by

$$\text{fBmFourierColor}[2, width, h, width, scale],$$

where *width* is the width (and height) of the pattern (in pixels), h is the Hurst exponent (between 0 and 1), giving a fractal dimension of $3 - h$ (between 2 and 3), and *scale* is an optional scaling of the absolute values. The higher the fractional dimension, the wilder the pattern will look. The generated matrix of colors can be written to a file (in ppm format) with

$$\text{WritePPM}[\text{"}file.\text{ppm"}, matrix, comment].$$

The optional comment is written into the file's header. The code is in the package fBm.m (see Listing 12.5–2).

```
fBmFourierColor::usage = "fBmFourierColor[e, n, h, maxn, scale] generates fBm
    color data using the Fourier synthesis method."
fBmFourierColor[2, n_?EvenQ, h_, maxn_:Automatic, scale_:1.0] :=
    Module[{four, full, i, j, n2 = n/2, ext, mbeta2 = -(h + 1)/2.0, nmax = maxn},
        If[ nmax === Automatic, nmax = n ];
        nmax = Round[nmax/2];
        four = Table[freq2[i, j, mbeta2, nmax], {i, 0, n2-1}, {j, 0, n2-1}];
        ext = Table[0, {n2}];
        four = Join[#, ext]& /@ four;
        ext = Table[0, {n}];
        full = Join[four, Table[ext, {n2}]];

        full = N[Sqrt[n]] InverseFourier[full];
        Map[colorGen, scale full, {2}]
    ]
colorGen[z_] := ToColor[ Hue[Arg[z]/(2Pi), 1, Mod[Abs[z], 1.0]], RGBColor ]
```

Listing 12.5–2: Part of fBm.m: Fourier synthesis for color data

The next example uses the tube from the ray-tracing example (Plate 11-b) in Section 11.4.4. Here are the commands used to generate the pattern and the tube in *Mathematica*.

We read the package for the fractal pattern generator.

```
In[13]:= Needs["MathProg`fBm`"]
```

For the color image, we generated a pattern 360 pixels wide. It has a fractal dimension $3-h = 2.7$. Here, we use a lower-resolution version.

```
In[14]:= (pat = fBmFourierColor[2,120,0.3,120,6.])//Short
Out[14]//Short=
    {{RGBColor[0., 0.435127, 0.161952], <<118>>,
        RGBColor[0., 0.568379, 0.291866]}, <<118>>, {<<120>>}}
```

Here is a grayscale version of this pattern.

```
In[15]:= Show[ Graphics[RasterArray[pat]],
                AspectRatio -> Automatic ];
```

12.5   Interface to External SIS Generators

We write it to a file, for use in RAYSIS. The string is written into the header of the output file (as documentation).

```
In[16]:= WritePPM["/tmp/pat360.ppm", pat,
               "fBmFourierColor[2, 360, 0.3, 360, 6.]"]
Out[16]= /tmp/pat360.ppm
```

The tube graphic is generated with these options. See Section 11.4.4 for the complete code needed to generate the graphic. Because the graphic is much smaller in the z direction than in the other two dimensions, it is best looked at straight down. Note the setting for the view vertical. Its default is {0,0,1}, which would be parallel to our direction of viewing. *Mathematica* would not complain about this degenerate case, but RAYSIS (rightly) does.

```
In[17]:= tube[{x5, y5}, {3, 3/2, 2}, PlotPoints -> {12, 60},
             ViewPoint -> {0,0,3}, ViewVertical -> {0,1,0}];
```

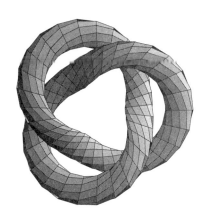

We write out the graphic in the form required by RAYSIS.

```
In[18]:= RaySis["/tmp/tube.sis", %]
Out[18]= /tmp/tube.sis
```

Plate 14-b was rendered with this command, using +P *file*.ppm to specify the pattern file:

```
raysis tube.sis -W 1770 -H 1200 +P pat360.ppm +o /tmp/tube.ppm
```

(The input file was edited to optimize the camera position.)

Plate 15 shows the diffraction pattern generated by two slits of finite width near the center of the bottom edge. The commands to produce the image data can be found in the notebook SIS-Examples.nb on the CD-ROM. The pattern used in RAYSIS was derived from a contour plot of $\sin x \cos(x/2 + y)^2$.

The CD-ROM contains the packages SIS.m and RaySis.m, as well as the notebook SIS-Examples.nb, with the examples from this chapter.

12.6 The Complete Code of SIS.m

```
BeginPackage["MathProg`SIS`"]

SIS::usage = "SIS[hxy, {x, x0, x1}, {y, y0, y1}, opts...] produces
    a SIS from the given function looking from the positive z axis."

SIRDS::usage = "SIRDS[-SurfaceGraphics-, opts...] produces a SIRDS from the given surface,
    contour, or density graphics, looking from  the positive z axis."

PlaneDistance::usage = "PlaneDistance -> 2 is the distance from the
    back plane to the viewing plane."

EyeDistance::usage = "EyeDistance -> 2 is the distance from the
    viewing plane to the eyes."

EyeSeparation::usage = "EyeSeparation -> 1/4 gives the separation of the
    eyes measured as a fraction of the horizontal width of the SIS."

Object::usage = "Object -> Point gives the object to draw for SIS. It is a function of
    {x, y}. The option is re-evaluated for each line."

Guides::usage = "Guides -> True draws guide dots at the top of a stereogram."

PlotPoints::usage = PlotPoints::usage <>
"PlotPoints -> 100 gives the number of random initial points used in SIS."

PlotRange::usage = PlotRange::usage <> "PlotRange -> {0, 1} gives the expected range
    of the function hxy over the entire range in SIS."

PlotStyle::usage = PlotStyle::usage <>
"PlotStyle -> {directives..} gives the graphics directives for
    rendering the objects in a SIS. Default is {PointSize[0.01]}."

Options[SIS] = {
    PlaneDistance -> 2,
    EyeDistance -> 2,
    EyeSeparation -> 1/4,
    Guides -> True,
    PlotPoints -> 100,
    Object -> Point,
    PlotRange -> {0, 1},
    PlotStyle -> {PointSize[0.01]}
};

Options[SIRDS] = {
    PlaneDistance -> 2,
    EyeDistance -> 2,
    EyeSeparation -> 1/4,
    Guides -> True
};

Begin["`Private`"]

Needs["Utilities`FilterOptions`"]
```

SIS.m

```
(* FindRoot Options to use *)
frOpts = Sequence @@ {AccuracyGoal -> 4}
SIS[ hxy0_, {x_, x0_, x1_}, {y_, y0_, y1_}, opts___ ] :=
    Module[{hxy, es0, es, d1, d2, ez, px, py, xx, pp, leye, reye, xmin = N[x0],
            xmax = N[x1], ymin=N[y0], ymax=N[y1], yeye,
            far, near, hmax, res, line, obj, guides, ps, next},
        es0 = EyeSeparation /. {opts} /. Options[SIS];
        es = es0 (xmax - xmin);
        {far, near} = PlotRange /. {opts} /. Options[SIS];
        hmax = near - far;     (* depth range *)
        hxy = hxy0 - far;      (* normalize *)
        pp = PlotPoints /. {opts} /. Options[SIS];
        leye = (xmin + xmax - es)/2.0;      (* left eye  *)
        reye = (xmin + xmax + es)/2.0;      (* right eye *)
        yeye = (ymin + ymax)/2.0;     (* eye y *)
        d1 = hmax PlaneDistance /. {opts} /. Options[SIS];
        d2 = hmax EyeDistance /. {opts} /. Options[SIS];
        ez = d1 + d2;
        ps = PlotStyle /. {opts} /. Options[SIS];
        If[ !ListQ[ps], ps = {ps} ];
        (* local function (shares many variables) *)
        next[xx_, yy_, eye1_, eye2_] :=
          Module[{xval, xlow, xhigh, sol, zval, zx, yx, h},
            {xlow, xhigh} = Sort[xx + {0.2, 1.2}d1(xx-eye1)/d2];
            (* z(x) and y(x) *)
            zx = Expand[d1 + (x-xx)d2/(eye1-xx)];
            yx = Expand[yeye + (ez-zx)(yy-yeye)/d2];
            (* f(x, y(x)) *)
            h = hxy /. y -> yx;

            sol = FindRoot[h == zx, {x, xlow, xhigh}, Evaluate[frOpts]];
            If[ !NumberQ[x /. sol], sol = {x -> (xlow + xhigh)/2} ];
            x + (d1-zx)(eye2-x)/(ez-zx) /. sol
          ];

        res = Table[ (* one point and its neighbors *)
            obj = Object /. {opts} /. Options[SIS]; (* eval object *)
            {px, py} = {Random[Real, {xmin, xmax}], Random[Real, {ymin, ymax}]};
            line = {obj[{px, py}]};
            (* extend to right *)
            xx = px;
            While[True,
                xx = next[xx, py, leye, reye];
                If[ xx >= xmax, Break[] ];
                AppendTo[line, obj[{xx, py}]];
            ];
            (* extend to left *)
            xx = px;
            While[True,
                xx = next[xx, py, reye, leye];
```

```
                    If[ xx <= xmin, Break[] ];
                    AppendTo[line, obj[{xx, py}]];
                ];
                line
            , {pp}];

        guides = If[ Guides /. {opts} /. Options[SIS],
            With[{xm = (xmin + xmax)/2, ym = ymin + 1.01(ymax-ymin)},
                {{PointSize[0.01], Point[{xm - es/4, ym}],
                  Point[{xm + es/4, ym}]}}
            ], {}, {} ];
        res = Join[ guides,
                {{Thickness[0],
                    Line[{{xmin, ymin}, {xmax, ymin},
                        {xmax, ymax}, {xmin, ymax}, {xmin, ymin}}] }},
                    ps, res ];
        Show[Graphics[ res,
            {PlotRange->Automatic, FilterOptions[Graphics, opts],
                AspectRatio -> Automatic}
        ]]
    ]

SIRDS[ gr:(SurfaceGraphics|ContourGraphics|DensityGraphics)[zmat_, __], opts___ ] :=
    Module[{es0, es, ez, d1, d2, nx, ny, hmax,
            mat, xmin, xmax, ymin, ymax, far, near, guides},
        es0 = EyeSeparation /. {opts} /. Options[SIRDS];
        {ny, nx} = Dimensions[zmat];
        {{xmin, xmax}, {ymin, ymax}, {far, near}} = FullOptions[ gr, PlotRange ];
        (* flat surface? *)
        If[ near - far < 10^-10, near = far + 10.0^-10 ];
        hmax = near - far;
        d1 = hmax PlaneDistance /. {opts} /. Options[SIS];
        d2 = hmax EyeDistance /. {opts} /. Options[SIS];
        ez = d1 + d2;
        es = Round[es0 nx];          (* scaled eye separation *)

    SIRDSLine[line_] :=
        Module[{n = Length[line], vals, i, j, r, left, dx},
            (* clip and normalize *)
            dx = Max[far, Min[near, #]]& /@ line - far;
            dx = es (d1-dx)/(ez-dx);  (* point distances *)
            left = Round[Range[n] - 0.5 dx];
            r = Dispatch[ Thread[left -> Round[left + dx]] ];
            vals = Table[ junk, {n} ];
            Do[ j = i /. r; vals[[i]] = If[ i < j <= n, vals[[j]], randDot ];
                , {i, n, 1, -1} ];
            vals
        ];

    mat = SIRDSLine /@ zmat;

    guides = If[ Guides /. {opts} /. Options[SIRDS],
        With[{ym = ny+1},
```

```
                    {{PointSize[0.01], Point[{nx(0.5 - es0/4), ym}],
                      Point[{nx(0.5 + es0/4), ym}]}}
                ], {}, {} ];
            Show[Graphics[ {guides, Raster[mat]},
                {FilterOptions[Graphics, opts], PlotRange -> All,
                 AspectRatio -> N[(ymax - ymin)(ny+1)/ny/(xmax - xmin)] }
            ]]
        ]
    ]
    randDot := Random[Integer]     (* 0 or 1 *)
End[]
Protect[Evaluate[$Context <> "*"]]
EndPackage[]
```

Listing 12.6–1: SIS.m

Appendices

References

[1] H. Abelson and G. J. Sussman. *Structure and Interpretation of Computer Programs*. The MIT Press, Cambridge, MA, 1985.

[2] Alfred V. Aho, Ravi Sethi, and Jeffrey D. Ullman. *Compilers: Principles, Techniques, and Tools*. Addison-Wesley, Reading, MA, 1986.

[3] R. Bailey. An Introduction to Functional Programming using HOPE. Technical Report 120, Imperial College, London, 1984.

[4] Dror Bar-Natan. Random-dot stereograms. *The Mathematica Journal*, 1(3):69–75, 1991.

[5] H. P. Barendregt. *The Lambda Calculus*. Studies in Logic 103. North Holland, Amsterdam, revised edition, 1984.

[6] Elwyn R. Berlekamp, John H. Conway, and Richard K. Guy. *Winning Ways for Your Mathematical Plays*. Academic Press, San Diego, CA, 1982.

[7] Allen H. Brady. The busy beaver game and the meaning of life. In Herken [28].

[8] Ivan Bratko. *Prolog Programming for Artificial Intelligence*. Addison-Wesley, Reading, MA, 1986.

[9] M. Brückner. *Vielecke und Vielflache*. 1900.

[10] F. W. Clocksin and C. S. Mellish. *Programming in Prolog*. Springer-Verlag, Berlin, 1981.

[11] H. S. M. Coxeter, Patrick du Val, H. T. Flather, and J. F. Petrie. *The Fifty-Nine Icosahedra*. Springer-Verlag, Berlin, 1982. Originally published by Univ. of Toronto Press, 1938.

[12] H. S. M. Coxeter, M. S. Longuet-Higgins, and J. C. P. Miller. Uniform polyhedra. *Phil. Trans. Royal Soc. London, Ser. A*, 246:401–450, 1953.

[13] Michael Crichton. *Jurassic Park*. Random House, New York, 1991.

[14] H. B. Curry. Grundlagen der kombinatorischen Logik. *Am. J. Math.*, 52:509–536,789–834, 1930.

[15] Martin Davis, Ron Sigal, and Elaine J. Weyuker. *Computability, Complexity, and Languages*. Academic Press, San Diego, CA, second edition, 1994.

[16] Robert L. Devaney. *An Introduction to Chaotic Dynamical Systems*. Benjamin/Cummings, 1986.

[17] A. K. Dewdney. Mathematical recreations. *Scientific American*, 252, April 1985.

[18] C. Henry Edwards. Twisted tubes. *The Mathematica Journal*, 3(1):10–13, 1993.

[19] M. J. Feigenbaum. Quantitative universality for a class of nonlinear transformations. *J. Stat. Phys.*, 19:25–52, 1978.

[20] Daniel P. Friedman and Matthias Felleisen. *The little LISPer*. The MIT Press, Cambridge, MA, 1987.

[21] Judith L. Gersting. *Mathematical Structures for Computer Science*. W. H. Freeman and Company, San Francisco, CA, second edition, 1987.

[22] Andrew S. Glassner, editor. *An Introduction to Ray Tracing*. Academic Press, San Diego, CA, 1991.

[23] Oliver Gloor, Beatrice Amrhein, and Roman E. Maeder. *Illustrierte Mathematik*. BirCom, Basel, 1994. CD-ROM mit Begleitheft.

[24] Oliver Gloor, Beatrice Amrhein, and Roman E. Maeder. *Illustrated Mathematics*. TELOS/Springer-Verlag, Santa Clara, CA, 1995. CD-ROM with booklet.

[25] Theodore Gray and Jerry Glynn. *Exploring Mathematics with Mathematica*. Addison-Wesley, Reading, MA, 1991.

[26] Zvi Har'El. Uniform solution for uniform polyhedra. *Geometriae Dedicata*, 47:57–110, 1993.

[27] Robert Harper, Robin Milner, and Mads Tofte. *The Definition of Standard ML*. The MIT Press, Cambridge, MA, 1990.

[28] Rolf Herken, editor. *The Universal Turing Machine: A Half Century Survey*. Kammerer & Unverzagt, Hamburg–Berlin, 1988.

[29] J. Roger Hindley and Jonathan P. Seldin. *Introduction to Combinators and λ-Calculus (London Math. Soc. Student Texts; 1)*. Cambridge Univ. Press, London, 1986.

[30] Seiji Horibuchi, editor. *Stereogram*. Cadence Books, San Francisco, CA, 1994.

[31] P. R. Hudak et al. Report on the programming language Haskell, a non-strict purely functional language, Version 1.2. *ACM SIGPLAN Notices*, 1992.

[32] *J. of Irreproducible Results*. Chicago Heights.

[33] Craig E. Kolb. *Rayshade User's Guide and Reference Manual*, January 10, 1992.

[34] Leslie Lamport. *LaTeX: A Document Preparation System*. Addison-Wesley, Reading, MA, 1986.

[35] Leslie Lamport. *MakeIndex: An Index Processor for LaTeX*, February 17, 1987.

[36] Roman E. Maeder. Mathematica as a programming language. *Dr. Dobbs Journal*, February 1992.

[37] Roman E. Maeder. *Informatik für Mathematiker und Naturwissenschaftler — Eine Einführung mit Mathematica*. Addison-Wesley (Germany), Bonn, 1993.

[38] Roman E. Maeder. *The Mathematica Programmer*. AP Professional, Boston, MA, 1994.

[39] Roman E. Maeder. Term rewriting and programming paradigms. In V. Keränen, editor, *Mathematics with a Vision: Proceedings of the First International Mathematica Symposium*, pages 7–19. Computational Mechanics Publications, 1995.

[40] Roman E. Maeder. *Programming in Mathematica*. Addison-Wesley, Reading, MA, third edition, 1996.

[41] Roman E. Maeder. Term Rewriting and Programming Paradigms. In Troels Petersen, editor, *Elements of Mathematica Programming*, TELOS/Springer-Verlag, Santa Clara, CA, 1996.

[42] John McCarthy. Recursive functions of symbolic expressions and their computation by machine I. *J. ACM*, 3:184–195, 1960.

[43] N.E. Thing Enterprises. *Magic Eye: A New Way of Looking at the World*. Andrews and McMeel, 1993.

[44] K.J. Paradise and John Fultz. Customizing tick marks. *The Mathematica Journal*, 5(2):42–44, 1995.

[45] Oren Patashnik. *BIBTEXing*, February 8, 1988.

[46] Heinz-Otto Peitgen, Hartmut Jürgens, and Dietmar Saupe. *Chaos and Fractals: New Frontiers of Science*. Springer-Verlag, Berlin, 1992.

[47] Heinz-Otto Peitgen and Dietmar Saupe. *The Science of Fractal Images*. Springer-Verlag, Berlin, 1988.

[48] POV-Ray Team. *Persistence of Vision Ray Tracer (POV-Ray) Version 2.0 User's Documentation*, 1993.

[49] P. Prusinkiewicz and A. Lindenmayer. *The Algorithmic Beauty of Plants*. Springer-Verlag, Berlin, 1990.

[50] T. Rado. On non-computable functions. *Bell Sys. Tech. J.*, pages 877–884, May 1962.

[51] Tomas Rokicki. *DVIPS: A TeX Driver*, 1993.

[52] M. Schönfinkel. Ueber die Bausteine der mathematischen Logik. *Math. Annalen*, 92:305–316, 1924.

[53] Robert Sedgewick. *Algorithms in C*. Addison-Wesley, Reading, MA, 1990.

References

[54] Steven S. Skiena. *Implementing Discrete Mathematics: Combinatorics and Graph Theory with Mathematica.* Addison-Wesley, Reading, MA, 1990.

[55] J. Skilling. Uniform polyhedra. *Phil. Trans. Royal Soc. London, Ser. A*, 278:111–135, 1975.

[56] George Springer and Daniel P. Friedman. *Scheme and the Art of Programming.* McGraw-Hill Book Company, New York, 1989.

[57] Alan M. Turing. On computable numbers with an application to the entscheidungsproblem. *P. Lond. Math. Soc. (2)*, 42:230–265, 1936-7.

[58] David A. Turner. An overview of Miranda. In David A. Turner, editor, *Research Topics in Functional Programming*. Addison Wesley, Reading, MA, 1990.

[59] A. H. Wheeler. Certain forms of the icosahedron and a method for deriving and designating higher polyhedra. In *Proc. Int. Math. Congress*, volume 1, pages 701–708, 1924.

[60] Stephen Wolfram. *Mathematica: A System for Doing Mathematics by Computer.* Addison-Wesley, second edition, 1991.

[61] Stephen Wolfram. *The Mathematica Book.* Wolfram Media, 3rd edition, 1996.

Index of Programs

This table lists the page numbers of all program listings. Note that some programs are not listed in a single listing, but have been spread out into several logical parts. A few others are not reproduced at all. The page refers to the place where they are mentioned.

Program	Page	Program	Page
Combinators.m	77	RayWalk.m	250
DAG.m	23	SIS.m	274
fBm.m	168	SortAux.m	121
FlipBookAnimation.m	133	SortAuxG.m	125
FSA.m	48	Sorting.m	122
Icosahedra.m	227	SortVisual.m	123
init.m	xiii	SphereWalk.m	4
IteratedFunctions.m	139	SphericalTriangles.m	206
JurassicPark.m	54	SurfaceGraphics3D.m	239
Lisp.m	44	Tape.m	93
Lists.m	45	Turing.m	94
LogicProgramming.m	33	TuringMacros.m	113
NIM.m	50	TuringOptimizer.m	108
Numerals.m	85	TuringRecursive.m	105
PolyhedraExamples.m	204	UniformPolyhedra.m	191
POVray.m	251	Unify.m	28
rayshade.m	243	Y.m	70
RaySis.m	269		

All programs (and those from the first volume [38]) are enclosed on *The Mathematica Programmer* CD-ROM. See page xv for details. Additionally, the CD-ROM contains the following packages, which are not listed in the book: BookPictures.m, init.m, and TuringExamples.m.

Index

Symbols beginning with a dollar sign $ are listed under the next letter. Names of people are given in italics. Further typographical conventions are explained on page xv.

& (Function), 65
&& (And), 34
' (Derivative), 61
//. (ReplaceRepeated), 81
: (pattern or default), 222
:- (PROLOG rule), 24, 53
:= (SetDelayed), 11
:> (RuleDelayed), 266
; (in PROLOG), 43
?- (PROLOG prompt), 41
@@ (Apply), 159
#0 (Slot[0]), 65
#ifdef, 101
{} (empty list), 63
` (context mark), 19, 77
| (Alternatives), 31
| (in PROLOG), 44
|| (Or), 35
⊥ (bottom), 69
⊔ (lub), 69
⊑ (partial order), 69

0, 98
59, 217
75, 189
68000 (Motorola), 108

Abstraction, 60, 75
   combinatory, 78–81
Acception
   of input, 47
   of language, 90
add, 60, 75
Addition
   in combinatory algebras, 86
   on Turing machine, 105
Additions, random, 167
addOne, 95

Adjacency, 199
Again, 25, **43**
Algebra, combinatory, 75–78
And, 31, 34, 52
Animations
   of sorting, 123–129
   static view, 132
Antiprisms, 187
Application
   of functions, 75
   self, 70
Apply, 159
Approximation
   by bisection, 146
   of functions, 69
   of initial conditions, 153
arc, 23
Archimedes, 187
Archive, on the WWW, xvi
args, 105
Arguments, 6
   several, 64
   unevaluated, 8
Arrays, 62
ASCII, 266
Assembler, 95
Assert, 31, **43**
assert, 39
Assignment, parallel, 121
Asymptotics, 130
Autoloading of packages, xiii
Automata (FSA), 47, 89
Axioms, 24, 52

B (combinator), 83
Backquote, `, 19, 77
Backtracking, 16, 29, 47–51
Beaver, busy, 110

BeginPackage, 19
BifurcationDiagram, 149
Bifurcations, 142–149
BinaryCurry, 64
Blank, 30
Block, 84, 127, 159
bottom ($\bot$), 69
Boxed, 205
Brontosaur, 53
Buffer, z, 268

C, 6, 7, 59, 91
C++, 7
C (combinator), 83
c (combinator), 80
Calculus, predicate, 52
Call, 28
Calling convention, 96, 100
Camera, 240
car, 45
Cases, 106
Catch, 197
cdr, 45
CD-ROM, xv, 285
Chains, 69
Chaos, 152–157
Characteristic, 199
Characters, 39
Charts, bar, 156
Chiral, 214
Church thesis, 98
Circle, great, 205
Classes, 18
Clause, 52
Clear, 41
Clipping, 262
Closure, transitive, 107
Clouds, 175, 179, 245
Code, dead, 107
Coloring, of several objects, 140
Colors, 220
Columns, 95
$Columns, 132
Combinatorica, 124
Combinators, 75
  instead of functions, 81
  Y, 70, 84

Commands, 5
  notation of, xv
Complete partial order (CPO), 69
Composition, 98, 102, 144
  of instructions, 96
Compounds, 217
Computability, 110
Computation
  accepting, 47
  efficient, 106
Conditional, 38, 85
Cone (Fig.), 261
Configuration, of Turing machines, 93
cons, 45
Constants, 16
Constructors, 16
Context mark, `, 19, 77
Contexts, 19
  and file names, 124
Control structures, 62
Convergence, fast, 157
Conversion, of data types, 241
Convexify, 201
CosA, 199
Cosine iteration (Fig.), 58
Coxeter, H. S. M., 189, 212
Craters, 173
Crichton, Michael, 53
Crystal, 245
Cube, 187
  snub, 191
Cuboids, 244
Currying, 64
Cut, 38
Cycles, 141, 144
  limit, 149

D, 61
δ, 159
DAG (directed acyclic graph), 23
Data types
  abstract, 16
  destructive, 91
  for surfaces, 237
  graphic, 258
Databases, 53, 197
Dead code, 107
Debugging, 43

Deduction, 51–55
Defaults, 221
Definitions, immediate, 20
Density, 199
DensityGraphics, 258
Depth first, 47
Derivative, 61
Determinism, 47
Diagram, final state, 149–152
Digon, 191
Dimension
　fractal, 168, 271
　of SIS, 263
　of tensor (rank), 169
Dinosaurs, 53
Directives, graphic, 140
Dirhombicosidodecahedron, great, 202
Disjunction, 52
Dispatch, 261
Dispersion, 141
Displacement, average, 165
Display, 238
DisplayFunction, 138, 205
Distribute, 11, 218
Distribution, normal, 166
Dodecadodecahedron, inverted snub (Fig.), Plate 5, Plate 13
Dodecahedra, stellated, 211
Dodecahedron, 187
　great, 188
Dodecicosidodecahedron, small (Fig.), Plate 5, Plate 13
Domain, 67
Doubling, of periods, 146
DownValues, 39
Dynamics, 149

Edge, 35
EdgeForm, 220
emptyTape, 92
Encapsulation, 19
Enumeration, of stellations, 217–219
Equal, 36
Equality, 36
Equations, 39
ETH, 250
Euclid, 9

Evaluation
　body of pure function, 159
　lazy, 14
　order of, 82
　partial, 106
　of queries, 28
Exchange, 121
Experiments, numerical, 154
Expert systems, 53
Exponent
　Hurst, 166
　spectral, 176
ExtraInformation, 243
EyeDistance, 262

F (combinator), 85
faceGraphics, 215
FaceList, 199
Faces, hemispherical, 193
Facets, 214
FaceTypes, 199
Factorial, 63
Facts, 24
fail, 37
$Failed, 26
Failure, 28
false, 37
Faults, random, 180
fBm, 166
fBmFault, 180
fBmFourier, 176, 178
fBmRA, 168
Feigenbaum constant, 152, 161
Feigenbaum diagrams, 149–152
　(Fig.), 136, 150
Fibonacci, 40
Figures, *see* Pictures
File names, xv
final, 48
Final-state diagrams, *see* Feigenbaum diagrams
FindBifurcation, 148
FindPeriod, 146
FindRoot, 146, 265
First, 63
First class, 59
Fixed points, 65–71, 84, 143
　functional for, 68
　least, 69

stability of, 142
FixedPoint, 62
Focal width, 240
Fold, 62, 170
FortranForm, 244
Fourier transform, 176
Fractal, 166
    (Fig.), 22
    landscapes, 170
    pattern for SIS, 271
FSA (finite state automaton), 47, 89
FullOptions, 243
Function, 59
Functional, 67
FunctionIteration, 58, 138
Functions, 5
    as combinators, 81
    computable, 85
    continuous, 69
    Curried, 80
    higher order, 61
    iteration of, 137–142
    local, 261, 265
    monotone, 69
    of two variables (Fig.), 256
    partial, 99
    pure, 92
    recursive (computable), 98–100
    recursively defined, 65
    total, 99
    as Turing tapes, 91
    without variables, 75

Games, 49
Gammas, 199
Gauss, C. F., 166
Gayley, Todd, 65
gcd, 9
GCD (greatest common divisor), 9
GIF, 172
Gödel numbering, 110
Golden Ratio, 7
Graphs, 23, 107
ground, 40
Groups, dihedral, 189, 213

Hacking, 109
HASKELL, 59

Head, 44
    of Turing machines, 89
HemisphericalQ, 199
Heptahedron, 193
HoldAll, 8
HoldFirst, 81, 121
Holding, of arguments, 7
HoldPattern, 82
Hooks, 124
HOPE, 59
Horn clauses, 52
Hue, 220
Hue, 140, 271
Hurst exponent, 271

I (combinator), 76
I, overloading, 77
Icon, of *Mathematica*, 224, 245
Icosahedra, 220
Icosahedra, stellated, 250
    (Fig.), 210, 236
Icosahedron, 187, 212–219
    faceted, 224, 245
    great, 211
    great (Fig.), 186, 225, Plate 5, Plate 6
    truncated, 245
Icosidodecahedron, great, 188, 250
    (Fig.), 269, Plate 4-a, Plate 5, Plate 13,
        Plate 14-a
Identity, 138
Implementation, 16, 19
Implication, 52
Index, 287–296
    of packages, 285
Inequality, 36
Inheritance, 18
initialConfiguration, 94
Input, 36
instruction, 89
Instructions, of Turing machines, 89
Integrate, 61
Interactive computing, 5
Interface, 19
Intermittency, 156–157
Interpolation, linear, 169
Inversion, 16
Islands, 173

Iteration, 137
   fixed-point, 66
   graphical, 137–141
   Newton, 158
   of functions of several arguments, 159
   simultaneous, 146, 159
   to find bifurcations, 146
   to find super-attractive points, 158
   vs. recursion, 63
`Iteration`, 144

Join, 45
`join`, 45

K (combinator), 76
Kepler–Poinsot polyhedra, (Fig.), 188
Kepler, Johannes, 188, 211
Knots, 247

λ^*, 78
λ-calculus, 75, 98
Lacunarity, 167
Lambda, 60
`LambdaStar`, 79
Landscapes, artificial, 170
Language
   machine, 95
   of *Mathematica*, 5
   styles of, 10
Laplace transform, 9
`LaplaceTransform`, 9
LaTeX, xvii
`left`, 96
Light sources, 240
Line, 138
Line, scan, 258
`lineGraphics`, 213
LISP, 39, 59
`list`, 45
Lists, 15
   in PROLOG, 44
   sorting of, 123
Literal, 52
`log`, 6
Logic
   combinatory, 76, 98
   fuzzy, 55
`LogicalExpand`, 14
`LogicValues`, 24, **43**

`LogisticsMap`, 137
Logo, of WRI, 224
Loops, 62, 97
lub (⊔), 69

μ-scheme, 99, 106
Macro, 95
`makeFib`, 71
Map, 13, 61
`MapIndexed`, 13, 140
Maps
   linear, 220
   logistics, 137, 153
Mark, 90
The Mathematica Journal, ix
MathSource, xiv
Matrices, 13
   adjacency, 194
   diagonal, 13
   symmetric, 14
`member`, 46
Membership, 46
Memory conservation, 129
`merge`, 106
Messages, 18
Methods, 17
`mFact`, 67
Miller-Freeman, ix
Minimization, 99
`$MinPrecision`, 159
MIRANDA, 59
Mixing, 154
ML, 59
Möbius strip, 192
Möbius triangles, 189
Modularization, 19
`Module`, 19, 197
Modus ponens, 52
Motion, Brownian, 165–167

N, 215
`name`, 40
`Needs`, 19, 124
Negation, 36
   inconsistency of, 85
`Nest`, 62, 94, 159
`NestList`, 138
`NextConfiguration`, 93

NIM, 49
No, 43
Nondeterminism, 47
noop, 96
NoSpy, 43
Not, 35, 52
Notation, xv
NumberedPolyhedron, 202
NumberOfEdges, 199
NumberOfFaces, 199
NumberOfFacesPerType, 199
NumberOfFaceTypes, 199
NumberOfVertices, 199
Numbers
 approximate, 159
 as combinators, 86
 complex, 167
 random, 11, 166, 258

Object, 17
Octahedra, compound of, 221
Octahedron, 187, 211
OnesidedQ, 199
Operations
 positional, 13
 structural, 11
Optimization, 63, 106–109
Options, 138
 instead of directives, 149
 passing on, 150
 with delayed values, 266
Or, 35, 52
Orbits, 137
 ergodic, 155
 periodic, 155
 super-attractive, 157–161
OrbitToPoints, 138
Order, 121
 partial ($\sqsubseteq$), 69
ordered, 45
Ordering, by size, 222
OrderOfGroup, 199
Output, 36
 suppression of graphic, 138
Overloading, 17

P (combinator), 85
p, 98, 101

Packages, 19
 index, 285
 loading, xiv
Pairing, 85
Parameters, 6, 20
 formal, 60
 input and output, 46
ParametricPlot3D, 237
Partition, 138
PASCAL, 6, 59
Path, 23
path, 23
Pattern
 diffraction, 273, Plate 15
 for SIS, 271
 matching, 8–10, 26
 optional, 221
Pattern, 30
pattern2Var, 30
PBM, x, 172, 245, 271
Pentagram, 187, 211
PermutationPlot, 123
Permutations, 123, 199
Phase, random, 167, 271
Pictures
 addition, 88
 cone, 261
 convex uniform polyhedra, 187
 diffraction pattern, Plate 15
 Feigenbaum diagrams, 136, 150
 fixed point of cosine, 58
 function, 256
 great icosidodecahedron, 269, Plate 4-a, Plate 5, Plate 13, Plate 14-a
 great icosahedron, 186, 225, Plate 5, Plate 6
 inverted snub dodecadodecahedron, Plate 5, Plate 13
 Kepler–Poinsot polyhedra, 188
 random walk, 4, 12, 15, 249, Plate 4-c
 Sierpinski carpet, 22
 Sierpinski sponge, Plate 16
 small dodecicosidodecahedron, Plate 5, Plate 13
 snub cube, 192
 sombrero, 266
 sorting, 120
 star polygons, 202
 stellated icosahedra, 210, 236
 sunflower, 74

Torus, 237
tube, 273, Plate 11-b, Plate 14-b
Turing machine, 88
Plane
　image, 257
　subdivision of, 213
PlaneColoring, 220
PlaneDistance, 262
Plato, 187
play, 50
Plot, 61
PlotPoints, 259
PlotRange, 269
PlotTuring, 94
Plumbing, 250
plus, 99, 109
Poinsot, 188, 211
Points
　bifurcation, 145
　Feigenbaum, 152
　of view, 249
　random on sphere, 167
PointSize, 149
Polygon, 200
Polygons, 244
　regular, 187
　rendering of nonconvex, 201
　star, 188
Polyhedra, 211
　uniform, 187–204
PolyhedronQ, 199
$Post, 124
POSTSCRIPT, xvii
POVray, 243
POVRAY, x, 172, 243, 269
$Pre, 124
pred, 86
Predecessor, 105
Predicates, 16, 33
　for matrices, 13
　instead of functions, 45
　for sorting, 132, 222
Prepend, 63
$PrePrint, 124
Print, 82
print, 37
Prisms, 187
Procedures, 5

Programming, 10–18
　assembly, 95–97
　declarative, 15
　dynamic, 40
　functional, 10–14, 59–64
　logic, 15–16, 23–56
　mathematical, 9
　object-oriented, 17–18
　procedural, 6
Programs
　external (ray tracing), 238
　external (SIS), 268
　self reproducing, 65
Projection, 98
　equirectangular, 183
　parallel, 264
PROLOG, 15, 23–56
Proofs, 52

Query, 32, **43**
QueryAll, 25, **43**
Quicksort, 121
　(Fig.), 120

Radiosity, 241
Radius, 194
RandomPoly, 6
randWalk, 165
Range, 128
Raster, 261
RawConfiguration, 199
Ray tracing, 172, 202, 226, 240–241, 269
RAYSHADE, x, 243
RAYSIS, x, 269
Reachability, 107
Receiver, 18
Recursion, 104
　primitive, 98
　vs. iteration, 63
$RecursionLimit, 71, 84
Reference, call by, 6
References, 281–283
Rejection, 47
Relocation, 96
Renaming, of variables, 60
Rendering
　of icosahedra, 219–223
　of polyhedra, 200–202

photorealistic, 240–241
    stereographic, 249, 257
ReplaceList, 16
ReplacePart, 138
Representation, 17
Reshape, 19
Resolution, 52
Rest, 63
Retract, 43
retract, 39
Retrograde, 192
Return, 28
reverse, 15
Rewrite rules, 5
right, 96
RISC, 108
RuleDelayed, 266
Rules, 8, 81
    inference, 23

S (combinator), 76
Satisfiability, 52
Scale, of SIS, 263
SCHEME, 59
Schwarz triangles, 189
Scoping, 19–20
Search
    binary, 147
    exhaustive, 49
Selectors, 16
Sensitivity, 153–154
Separation, of eyes, 257
Sequence, 44
Sequence, 218
Series, time, 141–142, 157
Shell, 247
Short, xiii
Show, 238
Sierpinski carpet (Fig.), 22
Sierpinski sponge (Fig.), Plate 16
silent, 48
Simulation, of Turing machines, 91–95
SIRDS (single-image random-dot stereogram), 257–264
SIS (single-image stereogram), 257
Slot, 65
Snub polyhedra, 191
SnubQ, 199

Soccer ball, 245
Socrates, 53
Software engineering, 18
Solve, 145, 220
Sombrero (Fig.), 266
Sort, 132
sort, 16
sortAnimation, 126
Sorting, 12, 16, 121–122, 199
    (Fig.), 120
    lists, 45
Space, 90
SPARC, xii, 108
Specification, 16
Sphere, random walk on, 4, 249
SphericalFaults, 182
splice, 97
SplitLine, 6
Sponge, Sierpinski, 248
Springer, ix
Spy, 43
sqrt, 100
squareList, 62
Stability, of fixed points, 143
Stack, 32, 59
States, 47
    final, 149
    internal, 32
Stellation, 211
    chiral, 226
Stereo images, 249–250
Stereograms, 256–273
    wall-eye, 257
    wallpaper, 250
Strategy, 50
Streams, 14
Style, of graphics, 141
Subroutines, 5
Subsets, 62
Substitution, 6, 60
Subtraction, 100
succ, 86, 98, 100
Success, 28
Successor, 98
Sum, 10
Sunflower (Fig.), 74
Super-attractive, 157
SuperAttractiveSeries, 159

SurfaceGraphics, 258
SurfaceGraphics3D, 238
Surfaces, 237–239
  minimal, 246
swap, 121
$swapAction, 124
Symbols, 197
  of Turing machines, 89
SymmetryGroup, 199
Syntax, xii

T (combinator), 85
Table, 238
Tail, 44
take, 49
Tape (Turing machine), 89, 91–92
TELOS, ix
Tensors, 166
Term rewriting, 8–10
Terminology, xv
Terms, combinatory, 76
Tetrahedra, compounds of, 221
Tetrahedron, 187, 211
TeX, xvii
Textures, 241
Theorems, 52
Theory of computation, 85
Thread, 261
Through, 79
Throw, 197
TIFF, xvii, 245
TimeSeries, 142
toCombinators, 81
Torus (Fig.), 237
TotalNumberOfFacesPerType, 199
traceLevel, 43
trans, 48
Transitions, 47
Transpose, 12, 138, 200
Transposition, 12, 169
Trees, 47
Triangles
  snub, 191
  spherical, 189, 205
Trott, Michael, 65
True, 31
true, 37
Truncation, 188

Tubes, 247
  (Fig.), 273, Plate 11-b, Plate 14-b
Tuning, 109
Turing machine, 89–111
  (Fig.), 88
TuringList, 94
TypeOfFaces, 199
Types, *see* Data types
Typewriter style, xv
Tyrannosaur, 53

Unequal, 36
Unevaluated, 79
Unification, 26–28, 53
Unify, 26
update, 92
Upvalues, 17, 31, 197, 238

ValenceOfVertices, 199
Value, call by, 6
var, 30, 40
Variables
  Gaussian, 166
  in patterns, 20, 24
  instance, 18
  local, 7, 19
  meta, xv
  random, 166
VAX, 108
Velociraptor, 53
Verbatim, 30
Version of *Mathematica*, xii
VertexConfiguration, 199
VertexCoordinates, 199
VertexList, 199
Vertices, enumeration, 194
ViewCenter, 249
Viewing, hints for stereograms, 259
ViewPoint, 249
Visualization
  Brownian motion, 165
  chaos, 162
  icosahedra, 211
  polyhedra, 187
  of sorting, 124
  stereoscopic, 249, 257
  surfaces, 237

W (combinator), 83

Walk, random, 250
  (Fig.), 4, 12, 15, 249, Plate 4-c
Wallpaper, 271
Waves, 245
win, 50
With, 66, 266
Workstations, graphic, 242
WWW, xv
Wythoff, 199

Wythoff symbol, 187, 190

xv, x

Y (combinator), 70, 84
Yes, 43

zero, 86, 100